Forest Conservation and Management

Forest Conservation and Management

Edited by
Russell Santiago

Larsen & Keller
www.larsen-keller.com

Forest Conservation and Management
Edited by Russell Santiago
ISBN: 978-1-63549-126-5 (Hardback)

Larsen & Keller
Published by Larsen and Keller Education,
5 Penn Plaza,
19th Floor,
New York, NY 10001, USA

Cataloging-in-Publication Data

Forest conservation and management / edited by Russell Santiago.
p. cm.
Includes bibliographical references and index.
ISBN 978-1-63549-126-5
1. Forest conservation. 2. Forest management. 3. Forests and forestry. I. Santiago, Russell.
SD411 .F67 2017
333.75--dc23

The publisher's policy is to use permanent paper from mills that operate a sustainable forestry policy. Furthermore, the publisher ensures that the text paper and cover boards used have met acceptable environmental accreditation standards.

Printed and bound in China.

For more information regarding Larsen and Keller Education and its products, please visit the publisher's website www.larsen-keller.com

Table of Contents

Preface

Deforestation has lead to many serious problems on Earth like global warming, pollution, soil erosion, etc. Thus, the conservation and sustainable use of forest and its resources is the need of this hour. Forest conservation refers to the practice of using forest resources in the optimum way so that we fulfill our requirements without jeopardizing that of the future generations. This book aims to provide essential information about this field. It presents the complex subject of forest conservation in the most comprehensible and easy to understand language. The topics covered in this extensive text deal with the core subjects of this field. It will be of great help to graduates and post-graduates in the fields of landscape planning, sustainable forest management and governance.

To facilitate a deeper understanding of the contents of this book a short introduction of every chapter is written below:

Chapter 1- Forest conservation is the management of forests. Sustainable forest management is concerned with the ecological, economic and socio- cultural aspects related to forests. This chapter is an introductory chapter which will introduce briefly all the significant aspects of forest conservation.

Chapter 2- Forest management manages the administration, economic and social aspect of forests. Some of the techniques involved in the process of forest management are timber extraction, cutting roads and preventing fire. Other aspects elucidated in this text are ecoforestry, community forestry, close to nature forestry and high conservation value forest.

Chapter 3- Techniques and approaches are an important component of any field of study. The techniques mentioned in this chapter are afforestation, silviculture, reforestation, selection cutting and forest protection. Afforestation is the process of growing a forest in any area where there was no tree cover earlier. The following chapter elucidates the various techniques and approaches of forest conservation.

Chapter 4- This chapter is a compilation of the various types of forests. Some of the forests that have been explained in the following section are taiga, temperate forests and tropical rainforest. The difference between temperate forests and rainforests is the difference between their climates and the animals that adapt in these forests. The section helps the readers in understanding all the types of forests and the differences between them.

Chapter 5- Deforestation is one of the biggest threats to forests; deforestation is the clearance of forests. The purpose behind this clearance can be industrial or it can be done with the intention of converting these forests into farms. Alternatively, the threats of forests are forest dieback, illegal logging, clearcutting and wildfire. The aspects explained in this chapter are of vital importance, and provide a better understanding of forests.

Chapter 6- Restoration ecology, conservation movement and conservation biology are the vital aspects of forest conservation. Restoration ecology is the practice of restoring destroyed ecosystems whereas the conservation movement is a social movement that pursues to safeguard the environment. This section elucidates the aspects and developments of forest conservation.

I would like to share the credit of this book with my editorial team who worked tirelessly on this book. I owe the completion of this book to the never-ending support of my family, who supported me throughout the project.

Editor

1

Introduction to Forest Conservation

Forest conservation is the management of forests. Sustainable forest management is concerned with the ecological, economic and socio- cultural aspects related to forests. This chapter is an introductory chapter which will introduce briefly all the significant aspects of forest conservation.

Sustainable forest management (SFM) is the management of forests according to the principles of sustainable development. Sustainable forest management has to keep the balance between three main pillars: ecological, economic and socio-cultural. Successfully achieving sustainable forest management will provide integrated benefits to all, ranging from safeguarding local livelihoods to protecting the biodiversity and ecosystems provided by forests, reducing rural poverty and mitigating some of the effects of climate change.

The "Forest Principles" adopted at The United Nations Conference on Environment and Development (UNCED) in Rio de Janeiro in 1992 captured the general international understanding of sustainable forest management at that time. A number of sets of criteria and indicators have since been developed to evaluate the achievement of SFM at the global, regional, country and management unit level. These were all attempts to codify and provide for independent assessment of the degree to which the broader objectives of sustainable forest management are being achieved in practice. In 2007, the United Nations General Assembly adopted the Non-Legally Binding Instrument on All Types of Forests. The instrument was the first of its kind, and reflected the strong international commitment to promote implementation of sustainable forest management through a new approach that brings all stakeholders together.

Definition

A definition of SFM was developed by the Ministerial Conference on the Protection of Forests in Europe (FOREST EUROPE), and has since been adopted by the Food and Agriculture Organization (FAO). It defines sustainable forest management as:

The stewardship and use of forests and forest lands in a way, and at a rate, that maintains their biodiversity, productivity, regeneration capacity, vitality and their potential to fulfill, now and in the future, relevant ecological, economic and social functions, at local, national, and global levels, and that does not cause damage to other ecosystems.

In simpler terms, the concept can be described as the attainment of balance – balance between society's increasing demands for forest products and benefits, and the preservation of forest health and diversity. This balance is critical to the survival of forests, and to the prosperity of forest-dependent communities.

For forest managers, sustainably managing a particular forest tract means determining, in a tangible way, how to use it today to ensure similar benefits, health and productivity in the future. Forest managers must assess and integrate a wide array of sometimes conflicting factors – commercial and non-commercial values, environmental considerations, community needs, even global impact – to produce sound forest plans. In most cases, forest managers develop their forest plans in consultation with citizens, businesses, organizations and other interested parties in and around the forest tract being managed. The tools and visualization have been recently evolving for better management practices.

In 2014, at the request of Member States, the Food and Agriculture Organization of the United Nations developed and launched the Sustainable Forest Management Toolbox, an online collection of tools, case studies and other resources to support countries implementing sustainable forest management.

Because forests and societies are in constant flux, the desired outcome of sustainable forest management is not a fixed one. What constitutes a sustainably managed forest will change over time as values held by the public change.

Criteria and Indicators

Deforestation of native rain forest in Rio de Janeiro City for extraction of clay for civil engineering (2009 picture). An example of non sustainable forest management.

Criteria and indicators are tools which can be used to conceptualise, evaluate and implement sustainable forest management. Criteria define and characterize the essential elements, as well as a set of conditions or processes, by which sustainable forest management may be assessed. Periodically measured indicators reveal the direction of change with respect to each criterion.

Criteria and indicators of sustainable forest management are widely used and many countries produce national reports that assess their progress toward sustainable forest management. There are nine international and regional criteria and indicators initiatives, which collectively involve more than 150 countries. Three of the more advanced initiatives are those of the Working Group on Criteria and Indicators for the Conservation and Sustainable Management of Temperate and Boreal Forests (also called the Montreal Process), Forest Europe, and the International Tropical Timber Organization. Countries who are members of the same initiative usually agree to produce reports at the same time and using the same indicators. Within countries, at the management unit level, efforts have also been directed at developing local level criteria and indicators of sustainable forest management. The Center for International Forestry Research, the International Model Forest Network and researchers at the University of British Columbia have developed a number of tools and techniques to help forest-dependent communities develop their own local level criteria and indicators. Criteria and Indicators also form the basis of third-party forest certification programs such as the Canadian Standards Association's Sustainable Forest Management Standards and the Sustainable Forestry Initiative Standard.

There appears to be growing international consensus on the key elements of sustainable forest management. Seven common thematic areas of sustainable forest management have emerged based on the criteria of the nine ongoing regional and international criteria and indicators initiatives. The seven thematic areas are:

- Extent of forest resources
- Biological diversity
- Forest health and vitality
- Productive functions and forest resources
- Protective functions of forest resources
- Socio-economic functions
- Legal, policy and institutional framework.

This consensus on common thematic areas (or criteria) effectively provides a common, implicit definition of sustainable forest management. The seven thematic areas were acknowledged by the international forest community at the fourth session of the United Nations Forum on Forests and the 16th session of the Committee on Forestry. These thematic areas have since been enshrined in the Non-Legally Binding Instrument on All Types of Forests as a reference framework for sustainable forest management to help achieve the purpose of the instrument.

On January 5, 2012, the Montreal Process, Forest Europe, the International Tropical Timber Organization, and the Food and Agriculture Organization of the United Nations, acknowledging the seven thematic areas, endorsed a joint statement of collaboration to improve global forest related data collection and reporting and avoiding the proliferation of monitoring requirements and associated reporting burdens.

Ecosystem Approach

The Ecosystem Approach has been prominent on the agenda of the Convention on Biological Diversity (CBD) since 1995 . The CBD definition of the Ecosystem Approach and a set of principles for its application were developed at an expert meeting in Malawi in 1995, known as the Malawi Principles. The definition, 12 principles and 5 points of "operational guidance" were adopted by the fifth Conference of Parties (COP5) in 2000. The CBD definition is as follows

The ecosystem approach is a strategy for the integrated management of land, water and living resources that promotes conservation and sustainable use in an equitable way. Application of the ecosystem approach will help to reach a balance of the three objectives of the Convention. An ecosystem approach is based on the application of appropriate scientific methodologies focused on levels of biological organization, which encompasses the essential structures, processes, functions and interactions among organisms and their environment. It recognizes that humans, with their cultural diversity, are an integral component of many ecosystems.

Sustainable forest management was recognized by parties to the Convention on Biological Diversity in 2004 (Decision VII/11 of COP7) to be a concrete means of applying the Ecosystem Approach to forest ecosystems. The two concepts, sustainable forest management and the ecosystem approach, aim at promoting conservation and management practices which are environmentally, socially and economically sustainable, and which generate and maintain benefits for both present and future generations. In Europe, the MCPFE and the Council for the Pan-European Biological and Landscape Diversity Strategy (PEBLDS) jointly recognized sustainable forest management to be consistent with the Ecosystem Approach in 2006.

Independent Certification

Growing environmental awareness and consumer demand for more socially responsible businesses helped third-party forest certification emerge in the 1990s as a credible tool for communicating the environmental and social performance of forest operations.

There are many potential users of certification, including: forest managers, scientists, policy makers, investors, environmental advocates, business consumers of wood and paper, and individuals.

With third-party forest certification, an independent organization develops standards of good forest management, and independent auditors issue certificates to forest operations that comply with those standards. Forest certification verifies that forests are well-managed—as defined by a particular standard—and chain-of-custody certification tracks wood and paper products from the certified forest through processing to the point of sale.

This rise of certification led to the emergence of several different systems throughout the world. As a result, there is no single accepted forest management standard worldwide, and each system takes a somewhat different approach in defining standards for sustainable forest management.

In its 2009–2010 Forest Products Annual Market Review United Nations Economic Commission for Europe/Food and Agriculture Organization stated: "Over the years, many of the issues that previously divided the (certification) systems have become much less distinct. The largest certification systems now generally have the same structural programmatic requirements. "

Third-party forest certification is an important tool for those seeking to ensure that the paper and wood products they purchase and use come from forests that are well-managed and legally harvested. Incorporating third-party certification into forest product procurement practices can be a centerpiece for comprehensive wood and paper policies that include factors such as the protection of sensitive forest values, thoughtful material selection and efficient use of products.

The ecolabel of the Forest Stewardship Council for certified sustainable forest product.

There are more than fifty certification standards worldwide, addressing the diversity of forest types and tenures. Globally, the two largest umbrella certification programs are:

- Programme for the Endorsement of Forest Certification (PEFC)
- Forest Stewardship Council (FSC)

The area of forest certified worldwide is growing slowly. PEFC is the world's largest forest certification system, with more than two-thirds of the total global certified area certified to its Sustainability Benchmarks.

In North America, there are three certification standards endorsed by PEFC – the Sustainable Forestry Initiative, the Canadian Standards Association's Sustainable Forest Management Standard, and the American Tree Farm System. FSC has five standards in North America – one in the United States and four in Canada.

While certification is intended as a tool to enhance forest management practices throughout the world, to date most certified forestry operations are located in Europe and North America. A significant barrier for many forest managers in developing countries is that they lack the capacity to undergo a certification audit and maintain operations to a certification standard.

Forest Governance

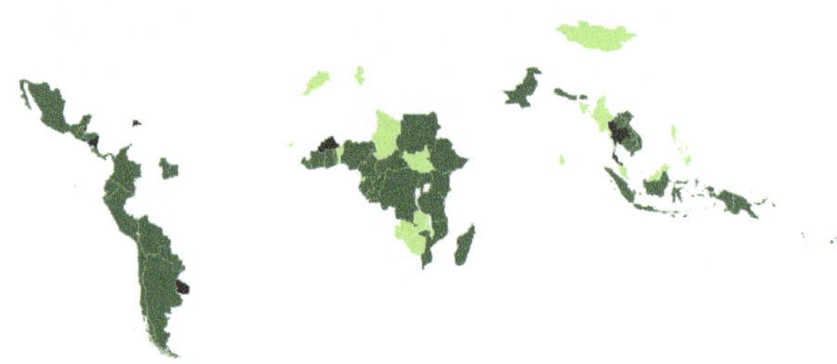

Countries participating in the UNREDD program and/or Forest Carbon Partnership Facility.

- UN-REDD participants
- Forest Carbon Partnership Facility participants
- participants in both

Although a majority of forests continue to be owned formally by government, the effectiveness of forest governance is increasingly independent of formal ownership. Since neo-liberal ideology in the 1980s and the emanation of the climate change challenges, evidence that the state is failing to effectively manage environmental resources has emerged. Under neo-liberal regimes in the developing countries, the role of the state has diminished and the market forces have increasingly taken over the dominant socio-economic role. Though the critiques of neo-liberal policies have maintained that market forces are not only inappropriate for sustaining the environment, but are in fact a major cause of environmental destruction. Hardin's tragedy of the common (1968) has shown that the people cannot be left to do as they wish with land or environmental resources. Thus, decentralization of management offers an alternative solution to forest governance.

The shifting of natural resource management responsibilities from central to state and local governments, where this is occurring, is usually a part of broader decentralization process. According to Rondinelli and Cheema (1983), there are four distinct decentralization options: these are: (i) Privatization – the transfer of authority from the central government to non-governmental sectors otherwise known as market-based service provision, (ii) Delegation – centrally nominated local authority, (iii) Devolution – transfer of power to locally acceptable authority and (iv) Deconcentration – the redistribution of authority from the central government to field delegations of the central government. The major key to effective decentralization is increased broad-based participation in local-public decision making. In 2000, the World Bank report reveals that local government knows the needs and desires of their constituents better than the national government, while at the same time, it is easier to hold local leaders accountable. From the study of West African tropical forest, it is argued that the downwardly accountable and/or representative authorities with meaningful discretional powers are the basic institutional element of decentralization that should lead to efficiency, development and equity. This collaborates with the World Bank report in 2000 which says that decentralization should improve resource allocation, efficiency, accountability and equity "by linking the cost and benefit of local services more closely".

Many reasons point to the advocacy of decentralization of forest. (i) Integrated rural development projects often fail because they are top-down projects that did not take local people's needs and desires into account. (ii) National government sometimes have legal authority over vast forest areas that they cannot control, thus, many protected area projects result in increased biodiversity loss and greater social conflict. Within the sphere of forest management, as state earlier, the most effective option of decentralization is "devolution"-the transfer of power to locally accountable authority. However, apprehension about local governments is not unfounded. They are often short of resources, may be staffed by people with low education and are sometimes captured by local elites who promote clientelist relation rather than democratic participation. Enters and Anderson (1999) point that the result of community-based projects intended to reverse the problems of past central approaches to conservation and development have also been discouraging.

Broadly speaking, the goal of forest conservation has historically not been met when, in contrast with land use changes; driven by demand for food, fuel and profit. It is necessary to recognize and advocate for better forest governance more strongly given the importance of forest in meeting basic human needs in the future and maintaining ecosystem and biodiversity as well as addressing climate change mitigation and adaptation goal. Such advocacy must be coupled with financial incentives for government of developing countries and greater governance role for local government, civil society, private sector and NGOs on behalf of the "communities".

National Forest Funds

The development of National Forest Funds is one way to address the issue of financing sustainable forest management. National forest funds (NFFs) are dedicated financing mechanisms managed by public institutions designed to support the conservation and sustainable use of forest resources. As of 2014, there are 70 NFFs operating globally.

References

- Matta, Rao (2015). Towards effective national forest funds, FAO Forestry Paper 174 (PDF). Rome, Italy: Food and Agriculture Organization of the United Nations. ISBN 978-92-5-108706-0.

- Matta, Rao (2015). Towards effective national forest funds, FAO Forestry Paper 174 (PDF). Rome, Italy: Food and Agriculture Organization of the United Nations. ISBN 978-92-5-108706-0.
- "Sustainable Forest Management Toolbox" (PDF). Food and Agriculture Organization of the United Nations. Retrieved 24 June 2014.
- "Evans, K., De Jong, W., and Cronkleton, P. (2008) "Future Scenarios as a Tool for Collaboration in Forest Communities". "S.A.P.I.EN.S." "'1'" (2)". Sapiens.revues.org. 1 October 2008. Retrieved 30 November 2011.
- "Mozgeris, G. (2008) "The continuous field view of representing forest geographically: from cartographic representation towards improved management planning". "S.A.P.I.EN.S." "'1'" (2)". Sapiens.revues.org. Retrieved 30 November 2011.
- "Pan-European Biological and Landscape Diversity Strategy". Strategyguide.org. Retrieved 30 November 2011.

2

Interdisciplinary Areas of Forest Conservation

Forest management manages the administration, economic and social aspect of forests. Some of the techniques involved in the process of forest management are timber extraction, cutting roads and preventing fire. Other aspects elucidated in this text are eco-forestry, community forestry, close to nature forestry and high conservation value forest.

Forest Management

Forest management is a branch of forestry concerned with overall administrative, economic, legal, and social aspects, as well as scientific and technical aspects, such as silviculture, protection, and forest regulation. This includes management for aesthetics, fish, recreation, urban values, water, wilderness, wildlife, wood products, forest genetic resources, and other forest resource values. Management can be based on conservation, economics, or a mixture of the two. Techniques include timber extraction, planting and replanting of various species, cutting roads and pathways through forests, and preventing fire.

Definition

The forest is a natural system that can supply different products and services. The working of this system is influenced by the natural environment: climate, topography, soil, etc., and also by human activity. The actions of humans in forests constitute forest management. In developed societies, this management tends to be elaborate and planned in order to achieve the objectives that are considered desirable.

Some forests have been and are managed to obtain traditional forest products such as firewood, fiber for paper, and timber, with little thinking for other products and services. Nevertheless, as a result of the progression of environmental awareness, management of forests for multiple use is becoming more common.

Public Input and Awareness

There has been increased public awareness of natural resource policy, including forest management. Public concern regarding forest management may have shifted from the extraction of timber for earning money for the economy, to the preservation of additional forest resources, including wildlife and old growth forest, protecting biodiversity, watershed management, and recreation. Increased environmental awareness may contribute to an increased public mistrust of forest management professionals. But it can also lead to greater understanding about what professionals do re forests for nature conservation and ecological services. The importance of taking care of the forests for ecological as well as economical sustainable reasons has been shown in the TV show Ax Men.

Many tools like GIS and photogrammetry modelling have been developed to improve forest inventory and management planning.

Wildlife Considerations

The abundance and diversity of birds, mammals, amphibians and other wildlife are affected by strategies and types of forest management.

Management Intensity

Forest management varies in intensity from a leave alone, natural situation to a highly intensive regime with silvicultural interventions. Management is generally increased in intensity to achieve either economic criteria (increased timber yields, non-timber forest products, ecosystem services) or ecological criteria (species recovery, fostering of rare species, carbon sequestration).

Criteria and Indicators of Sustainable Forest Management

Criteria & Indicators of Sustainable Forest Management (C&I) are policy instruments by which sustainability of forest management in the country/region, or progress towards Sustainable forest management (SFM), may be evaluated and reported on. C&I is a conjunctive term for a set of objectives and the variables/descriptions allowing to evaluate whether the objectives are achieved or not.

There are many various sets of C&I in the world that are used by particular regional SFM processes (e.g. FOREST EUROPE, Montreal Process), international organisations and their activities (e.g. FAO Global Forest Resources Assessment) or certification of forest management and forest products (e.g. Forest Stewardship Council, Programme for the Endorsement of Forest Certification). Signatory countries of particular processes or certification schemes can develop their national sets derived from the set of process/scheme.

Possible Uses of Criteria and Indicators

Criteria and indicators of sustainable forest management:

- help to define, understand and promote the concept of sustainable forest management;
- provide a common framework for signatory countries to:
    - o describe, monitor, assess and report on national forest trends (if measured periodically);
    - o describe, monitor, assess and report on progress towards sustainable forest management (if measured periodically);
- reflect a holistic approach to forests as ecosystems, highlighting the full range of forest values;
- facilitate policy dialogue and the development of policies or strategies ;
- help to implement forest related policies, plans and programmes;
- contribute to cross-sectoral sustainability assessments, as well as assessments for other sectors (e.g. environment, energy, climate change, agriculture, sustainable land management);
- guide forest management practice;
- help to identify the changes in forest management;
- help to develop forest certification principles, standards and indicators.

Levels of Use

Criteria and indicators are applied at different levels:

- global - the sets of C&I for global use have to be universal enough for the various biogeographic regions, countries at different stages of development of forestry (with different ability to provide data), etc;
- regional (supranational regions, such as pan-European, are meant);
- national and sub-national;
- local (forest management units, community forestry land, concession areas and other) - as being foccused on much smaller scale, the sets for this level can signifficantly differ from sets for the above-mentioned levels and they are used mainly for forest certification purposes.

Early History

The first FAO Global Forest Resources Assessment was published in 1948. It comprised five indicators, however, their main focus was to assess the availability of timber, not sustainability of forest management as a whole.

The history of the idea of C&I of SFM goes back to 1992 when the United Nations Conference on Environment and Development adopted the "Forest Principles"* and Chapter 11 of Agenda 21. At about the same time, the International Tropical Timber Organization (ITTO) did some pioneering work on "Criteria for the Measurement of Sustainable Tropical Forest Management." Following this summit, the concept of "criteria and indicators for sustainable forest management" gained increasing international attention.

Forest Europe: Pan-european Criteria and Indicators for Sustainable Forest Management

The first set of Pan-European C&I, based on documents adopted by two Expert Level Follow-Up Meetings of the Helsinki Conference, was adopted by the third Ministerial Conference on the Protection of Forests in Europe on 2–4 June 1998 in Lisbon/Portugal as an Annex 1 of the Resolution L2. This set consisted of 6 Criteria of SFM, 20 quantitative indicators and 80 descriptive indicators (4 per each quantitative indicator).

In the meantime, knowledge and data collection systems as well as information needs have gradually developed further. Thus, initiated through the Lisbon Conference, the MCPFE decided to improve the existing set. Document named Improved Pan-European Indicators for Sustainable Forest Management was adopted at expert level at the MCPFE Expert Level Meeting, 7–8 October 2002 in Vienna, Austria (before Vienna Conference). The 6 criteria remained unchanged, however, the number of quantitative indicators was increased to 35, and the system of descriptive indicators, renamed to qualitative indicators, was significantly simplified to 17 indicators.

The last update of Pan-European set was performed before Madrid Conference at the "Forest Europe" Expert Level Meeting 30 June – 2 July 2015, in Madrid, Spain. The set of quantitative indicators was slightly altered (34 indicators); the system of qualitative indicators was further simplified to 11 indicators, 5 of them now forming something like an "unofficial 7th criterion", while the remaining 6 are the official 6 criteria.

Criteria

Criteria characterise or define the essential elements or set of conditions or processes by which sustainable forest management may be assessed (MCPFE, 1998b). There are 6 criteria in the Pan-European set:

1. Maintenance and Appropriate Enhancement of Forest Resources and their Contribution to Global Carbon Cycles
2. Maintenance of Forest Ecosystem Health and Vitality
3. Maintenance and Encouragement of Productive Functions of Forests (Wood and Non-Wood)

4. Maintenance, Conservation and Appropriate Enhancement of Biological Diversity in Forest Ecosystems

5. Maintenance and Appropriate Enhancement of Protective Functions in Forest Management (notably soil and water)

6. Maintenance of other socioeconomic functions and conditions

Indicators

The indicators show changes over time for each criterion and demonstrate the progress made towards its specified objective (MCPFE, 1998a).

- Quantitative indicators are expressed in measurement units and the necessary date are collected via regular forest inventories, other field surveys, remote sensing, etc. Periodically measured indicators reveal the direction of change with respect to criterion. The list of quantitative indicators includes, for example, the forest area and growing stock (volume of living wood) for the Criterion 1, forest damage for the Criterion 2, increment and fellings for the Criterion 3, deadwood volume or naturalness classes for the Criterion 4, the area of protective forests for the Criterion 5, and contribution of forests to GDP or the area of recreation forests for the Criterion 6.

- Qualitative indicators have to be described and assessed and the data are collected using questionnaires. They are used to describe legal and institutional frameworks of forestry, as well as policies and istruments for the implementation of SFM.

Montreal Process

The Montreal Process (MP) Working Group on C&I for the Conservation and Sustainable Management of Temperate and Boreal Forests was launched in 1994 as a response to the Rio Forest Principles. Original set of C&I was adopted by Santiago Declaration in 1995. It consited of 7 criteria and 67 indicators. Current set represents the 5th version of MP C&I and it has 7 criteria and 54 indicators (both qualitative and quantitative.

Criteria

1. Conservation of biological diversity

2. Maintenance of productive capacity of forest ecosystems

3. Maintenance of forest ecosystem health and vitality

4. Conservation and maintenance of soil and water resources

5. Maintenance of forest contribution to global carbon cycles

6. Maintenance and enhancement of long-term multiple socio- economic benefits to meet # the needs of societies
7. Legal, institutional and economic framework for forest conservation and sustainable management

Ecoforestry

Ecoforestry has been defined as selection forestry or restoration forestry. The main idea of Ecoforestry is to maintain or restore the forest to standards where the forest may still be harvested for products on a sustainable basis. Ecoforestry is forestry that emphasizes holistic practices which strive to protect and restore ecosystems rather than maximize economic productivity. Sustainability of the forest also comes with uncertainties. There are other factors that may affect the forest furthermore than that of the harvesting. There are internal conditions such as effects of soil compaction, tree damage, disease, fire, and blow down that also directly affect the ecosystem. These factors have to be taken into account when determining the sustainability of a forest. If these factors are added to the harvesting and production that comes out of the forest, then the forest will become less likely to survive, and will then become less sustainable.

Since the forest is considered an ecosystem, it is dependent on all of the living and non-living factors within itself. This is a major part of why the forest needs to be sustainable before it is harvested. For example, a tree would convert sunlight to sugars for respiration to keep the tree alive. The remains of the converted sugars is left in roots for consumption by the organisms surrounding the tree in the habitat. This shows the productivity of an ecosystem with its inhabitants. Productivity within the ecosystem cannot come to fruition unless the forest is sustainable enough to be harvested. If all of the organisms of the ecosystem are non existent, the ecosystem itself will soon come to be non existent. Once that happens, there is no longer any forest to harvest from. The overall productivity of a system can be found in an equation where the Net Primary Production, or NPP, is equal to the Gross Primary Production, or GPP, minus the Respiration, or R. The formula is the NPP = GPP - R. The NPP is the overall efficiency of the plants in the ecosystem. Through having a constant efficiency in NPP, the ecosystem is then more sustainable. The GPP refers to the rate of energy stored by photosynthesis in plants. The R refers to the maintenance and reproduction of plants from the energy expended.

Ecoforestry has many principles within the existence of itself. It covers sustainable development and the fair harvesting of the organisms living within the forest ecosystem. There have been many proposals of principles outlined for ecoforestry. They are covered over books, articles, and environmental agencies. All of the principles relate to the idea that in ecoforestry, less should be harvested, and diversity must be managed. Through harvesting less, there is enough biomass left in the forest, so that the forest may stay healthy and still stay maintained. It will grow at a sustainable level annually, and thus it will be able to still be harvested the following year. Through management of the diversity, species may cohabitate in an ecosystem where the forest may feed off of other species in its growth and production. The Principles of Ecoforestry may be found below.

Principles of Ecoforestry

The principles of ecoforestry are:

Number	Principle
1	One must consider retention. It must be the first consideration in any planned removal of trees from a stand. By placing retention at the beginning of a planned removal, one makes sure that they know what must be left to ensure the protection of such things as rare species, sites of native cultural significance, riparian zones.
2	Riparian zones should not be touched. It is a sensitive area, thus tree removal should not occur there. This protects the water quality. Water quality is protected by not altering the drainage pattern of the zones. No tree removal should take place in the most sensitive areas.
3	The composition and structures should be upheld so that forests may fully function. This may be subject to large old trees, snags, and large fallen trees. These pieces of the ecosystem are upheld through letting them grow and die out into a timber extraction area.
4	When removing trees, the lowest impact should be used. This means not compacting the soil in the forest and building small or no roads if at all possible. This takes away the disturbance that would be done to the foundation of the forest's grounds.
5	Plan in terms of the needs of the larger watershed. The watershed zone plan designates areas where tree removal is not permitted. It also designates the areas where removal is possible, and the different types of removal available.
6	Prohibit clear cutting. Clear cutting is not ecologically correct. There are methods for cutting trees to maintain structure, and to allow the trees to cohabitate in a healthy ecosystem.
7	Select trees as candidates for removal by considering how abundant and redundant their structures and functions are to the rest of the forest as a whole, leaving potential wildlife trees (to become snags and coarse woody debris).
8	The forest should be allowed to regenerate trees through the seeds from the trees in the logged areas. This allows tree planting to be taken out of the ecoforestry principles, allowing a natural regeneration.
9	Ecological succession should be kept at all times. This will protect biological diversity. This even means eliminating the process of brush control.
10	No slash burning. Fire may be used as a tool in landscapes that have a history of naturally occurring fires. Although this may be true, fire should still be used with caution.

11	No pesticide use. The forest needs disease, insects, and shrub/herb vegetation. They are essential parts of a fully functioning forest, even though they involve the forest's decay in some cases. They are natural parts that allow the ecosystem to operate as a whole.
12	Maintain and restore topsoil quality. This can be done through leaving sufficiently coarse and small debris on the forest grounds.
13	Maintain beauty and other natural aesthetic qualities in the visual, sound, and odor landscapes. This includes not taking away the wildlife, the plants, or structural layout of the ecosystem.
14	Always look at the forest as a whole. Each part of the forest contributes to its overall needs and health. That is how the forest would have survived without human interference.
15	Rely more on people and markets. Use accounting and budgeting as a solution to rely less on the destruction or harvesting from the forests.
16	Don't do wrong. If it feels wrong, then it is probably wrong. Don't allow ignorance to persevere. Recognize that the ecosystem needs tending to. If the forest is not preserved, then it cannot be harvested forever.

Community Forestry

Community forestry is an evolving branch of forestry whereby the local community plays a significant role in forest management and land use decision making by themselves in the facilitating support of government as well as change agents. It involves the participation and collaboration of various stakeholders including community, government and non-government organisations (NGO's). The level of involvement of each of these groups is dependent on the specific community forest project, the management system in use and the region. It gained prominence in the mid-1970s and examples of community forestry can now be seen in many countries including Nepal, Indonesia, Korea, Brazil, India and North America.

National Forest Inventory of Peru

Overview

Community forestry is a branch of forestry that deals with the communal management of forests for generating income from timber and non-timber forest products as forms of goods while in other hand regulating ecosystem, downstream settlements benefits from watershed conservation, carbon sequestration and aesthetic values as in forms of services . It has been considered one of the most promising options of combining forest conservation with rural development and community empowerment and poverty reduction objectives. Community forestry is defined by the Food and Agricultural Organization of the United Nations as "any situation that intimately involves local people in forestry activity". Community forestry exists when the local community in an area plays a significant role in land use decision-making and when the community is satisfied with its involvement and benefits from the management of the surrounding forest and its resources.

Community forestry is first implemented through the establishment of a legal and institutional framework including the revision of legal norms and regulations for forest management, the development of National Forest Plans and the strengthening of decentralization processes to sub-national levels of government. The second principal line of action is the implementation of pilot projects to demonstrate the feasibility of the community forestry framework . However, a study by the Overseas Development Institute shows that the technical, managerial and financial requirements stipulated by the framework are often incompatible with local realities and interests. A successful legal and institutional framework will incorporate the strengthening of existing institutions and enable the dissemination of locally appropriate practices as well as the local capacity for regulation and control.

In a 2016 review of community-based forestry, FAO estimated that almost one-third of the world's forest area is under some form of community-based management.

History

Community forestry first came to prominence in the mid-1970s and has continued to evolve over the last few decades in a growing number of countries. The availability of forest resources are often greatly reduced for use by the local people due to increasing pressures to cultivate the land, reliance on the forest resources are also affected by economic and political changes. The evolution of community forestry in Nepal dates back to the late 1970s and was first instilled as an attempt to improve the management of forest resources and address environmental issues that were of great concern with the countries failing centralized forest policy. Over the past two decades, community forestry has been applied successfully in many developing countries, with its main goal being the alleviation of poverty amongst local forest communities and forest conservation. More recently, community forestry has been implemented in developing countries and it has been successful in its aims of sustainable forest management, climate change adaptation plan of action, and securing socio-economic benefits for local communities.

Stakeholders

There is a large variety of stakeholders involved when considering community forestry. Participation from some of the various levels of community, government and non-government organisations (NGO's) are essential in the project's success. While specific stakeholders vary between different community forestry projects the primary stakeholder groups are as follows:

Local Community
Communities living adjacent to or within the forest
Traditional authorities
Community-based organisations (forest users groups etc.)
Community representatives / local councils
Government
State Government
National Government
Departments of Agriculture, Forestry, Environment etc. (specific the country and/or region)
NGO's
Environmental/conservation groups
Commercial forestry industries
Industries reliant on forests (i.e. harvestable products other than wood, e.g. game meat)
Tourism industry
Animal welfare groups
Natural Ecosystem

Table 1. Common stakeholders involved with community forestry.

Stakeholders of community forestry have a vested interest to establish sustainable practices, whether this be to develop and maintain a regular income, ensure that forests are sufficiently protected to ensure their longevity or to reduce illegal activities and manage the area in such a way to promote tourism and conservation. In this situation, stakeholders came to conclusion to handover forest resources to local communities for conserving, managing and utilization by their own decision. Despite significant development, continued improvement in the collaboration between local governments and forest communities seems to be a key point for better community forest management. A wide range of futures scenarios have been put up to help the environmental decision process.

Challenges

A study conducted in the Brazilian Amazon determined that there are a number of challenges that must be faced when developing a sustainable management strategy for community forestry. These challenges are outlined in Table 2 shows the impact each management challenge has on other obstacles. The model is segregated into two phases: the *development phase* during which several enabling factors (land ownership, organisational capacity, technical knowledge and capital) are needed to obtain a legal management permit and secondly the *operational phase* where factors (clandestine loggers, access to markets, infrastructure and managerial skills) influence the successfulness of the management program. Each of the challenges outlined in Table 1 must be addressed in order for a self-sustaining community forestry management program to be established.

Land ownership	It is not uncommon for small settlements living traditionally to lack clear title to the forests and lands, as such their rights to harvest the land may come into dispute. Formal recognition of land ownership or rights to use is needed for legal forest management. In Nepal, there is still confusion to communities about their tenure into services such as income from trading of carbon sequestration come from forestry resources. Government wants to hold revenue raised from trading of carbon but communities claim that it is our property due tobe managed by us.
Organizational capacity	Organizational capacity refers to the community's competence to organize a forest management program.
Capital	A community will require start-up capital to invest in the required infrastructure, equipment, and to hire a forester to organise and oversee management plans.
Technical knowledge	While it is common for community members have a thorough understanding of forest ecology in a natural and historical sense, they often lack the technical knowledge and legal certification required to manage forest resources legally.
Legal management	Community forestry needs to be based on a legal management plan, prepared and approved by the relevant government authority (usually state environment agencies). Approval of the management plan can often be a long, bureaucratic process.
Clandestine loggers	Clandestine (illegal) loggers can enter a region and illegally log valuable species without the community's consent, and without a legal management plan. Illegal harvesting often significantly degrades the forest as few limitations are observed by the illegal loggers. Communities are either robbed of their valuable timber, or are paid below market prices.
Market access	Small villages that are often the focus of community forestry initiatives generally have limited access to markets due to their physical isolation, precarious transport and communication, limited contact with buyers and lack of marketing knowledge. Often they will also face difficulties competing with large-scale operations and illegal timber that flood the market with cheaper products.
Infrastructure	Forest management, for all types of enterprise, requires a certain amount of physical infrastructure, such as roads, logging equipment, buildings for storage and management headquarters and/or a reliable power supply.

Managerial skills	Skills related to effective management of the program, business knowhow and entrepreneurial ability, day-to-day decision-making, marketing skills, ability to resolve internal conflicts, and ensuring community benefit sharing are often lacked in rural villages.
Economic returns	The result of inadequacies in management is small and often insufficient economic returns to keep the program afloat, and furthermore to keep the community interested in the activity.

Table 2. Challenges faced by stakeholders managing forests

Best Practices

For a conservation program within a Community forest to succeed, community involvement is imperative. Governments with interest in forest conservation introduce statewide policies and legislations which have historically failed to deliver the desired outcomes such as in China, Nepal and Peru. Moreover, no single stakeholder by itself can ensure the success of such a program.

Community Forestry and wildlife Conservation

Though there is little research on the role of community forestry to wildlife conservation, some empirical studies suggests that it help in wildife conservation. It is done by decreasing the human disturbance, increasing regeneration of forest and increasing of ground cover.

Nepal

Common land in Nepal is owned by the state which often does not have the capacity to monitor and manage the area. This often leads to the over use of the resources by the community due to lack of incentives. To overcome this, programs involving community participation were introduced and 'Forest User Groups' (FUG) formed to manage the forests resources without giving them ownership of the land. Community forest management system in Nepal becomes one of the successful program out of 8 around world that is recognized on Rio 20+. This has resulted in better outcomes in the region.

Indonesia

Villagers from three communities in Bantaeng district in Indonesia, with the assistance of a funded project obtained a Forest management license and secured a 35-year lease on their local forest. With the incentive to preserve their source of income, the villagers have had a positive impact on forest conservation. This is now viewed as a reference model for the Ministry of Forestry's future projects.

Korea

Farmers enter into a profit sharing contract with the 'Village Forestry Association'

(VFA) to form a cooperative which assists farmers with reforestation in keeping with the legislation. VFA, though loosely linked to the Office of Forestry enjoys a degree of autonomy facilitating community participation. This system demonstrates the desirable mix of top-down and bottom-up planning ensuring government control as well as effective reforestation through active community participation.

Management System

As it is unlikely for any two areas to have identical socio-economic and environmental conditions, there is no blueprint management plan that would provide a blanket solution to all the problems. Based on research over several years in Nepal, it was noted that to have and effective management system, we need to identify variables which would affect the success of the system and group them into the following five sets:

(I) attributes of the resource system,

(II) attributes of the user group,

(III) attributes of the governance system,

(IV) attributes relating to interactions between the user group and resource, and

(V) attributes relating to interactions between the governance system and the resource.

In some cases, it is unrealistic to expect progress in a community level management of forest activities, as often conflict arise with respect to land use and benefit sharing within the community. Such issues can be overcome by recognising that a community level of management may not be the most effective management technique. Instead, adopting the following approach will provide a solution -

(I) Management by smaller work groups within the community, sharing common interest in the resources.

(II) A clear management plan with specific benefit sharing arrangement within the work group.

(III) Develop management systems which are within the expertise of the working group.

Faith communities are increasingly participating in efforts to promote ecological sustainability. Whereas the last 50–100 years has seen them lease out their territory to industry, they are beginning to reclaim and restore this land. Their recognition amongst local and national authorities have allowed community forestry schemes to develop and there have been several highly successful efforts across Cambodia, which have been extended to Vietnam and Laos. In September 2010, Buddhists monks were awarded the UNDP's Equator Prize for their ongoing conservation work. This comprises the establishment of tree nurseries, seedling distribution, ordination activities, composting schemes and a vegetable garden. They are using this work to promote sustainable living and conservation throughout three districts; the pilot project, launched in 1997 in Kratie consists of a six pagoda network that covers 1,461 hectares (3,610 acres). The site has become a valuable source of information and environmental education as well as a base for the local community. Villagers regularly hold panel discussions about how they can go on to best take advantage of what the natural forest has to offer without destroying it.

Incentives

Farm Forest for Household Use

(I) Preserving the source of community's livelihood

(II) Maintaining soil integrity by preventing erosion.

(III) Use of fuel wood and fodder- Planting trees to provide fuel wood as well as fodder for cattle can be encouraged in ways that do not conflict with cash crops and food production. e.g. The Neem tree was introduced in West Africa and is now the most widely grown tree in the drier parts of the continent. It was easily cultivated and provides the farmers with good timber, fuel and shade.

(IV) Developing nursery networks through support activities which assist with subsidies. This encourages entrepreneurship to produce tree planting stock for sale.

Farm Forests for Markets

Tree growing can take on the characteristics of a crop where there is a market for wood products such as poles, fuel wood and, pulp for production of paper. Companies tie up with farmers for supply of these products giving a steady source of income to the farming community. For example, in Philippines, over 3000 farmers cultivate trees for pulp production for an industry that provides a market as well as a minimum price for the product, and Cooperatives of Village Forestry Association in Korea have helped local communities cater to a growing market in forest products such as timber and mushrooms.

Close to Nature Forestry

Close to nature forestry is a theory and practice that takes the forest as an ecosystem and manages it as such. It is based on reduced human intervention, that should be directed to accelerate the processes that nature would do by itself more slowly. It aims at overcoming the divorce between forestalist and ecologist management systems of forest. As an important consequence, it concludes that if properly applied, it would render the segregation of forest lands into "productive" and "reserves" or national parks unnecessary.

Mixed and irregular deciduous forest in Catalonia

History

Europe

The Arbeitsgemeinschaft für Naturgemässe Waldwirtschaft (ANW) (German: *Working Group for the Close to Nature Forestry*) was established in Germany in 1950. In recent years this association has increased a lot its membership. The main reasons being the increase of ecological consciousness, the growing demand for forest products or services other than wood, the damages suffered by regular forest stands, the forest death fear. Another term often used to describe a close to nature approach to forest management, especially in Britain and Ireland, is Continuous Cover Forestry.

Because of a 1948 forest law, Slovenia has many forests managed according to the principles of close to nature forestry. In 1989 ANW promoted a meeting at Robanov Kot, in the Julian Alps, and the ProSilva organization was created, with representatives of 10 countries. At present the organization headquarters are in the French region of Alsace.

North America

In the United States professor Thom McEvoy has published the book *Positive Impact Forestry*, that recommends forestry practices similar to those of the "close to nature" movement. He thinks that the precursors of this type of forestry are to be found in Europe, mainly in Germany, and particularly makes mention of Heinrich Cotta, and his famous *Cotta's Preface*, that highlighted the importance that the study and understanding of nature should have for the foresters. As a more immediate precursor he makes reference to American forester and ecologist Aldo Leopold.

The Ecoforestry Institute consists on educational, non profit and non governmental organizations operating in US and Canada. They propose a forestry based on ecological principles, very similar to those of Pro Silva.

Forest Management/ecosystem Management

The *close to nature* approach intends to bridge the discrepancies, or even antagonisms between the *silvicultural* and *ecological* visions on the single reality of forest, considering the forest as an ecological system that produces wood. The sought after solution is not to segregate the territory into areas devoted to either forestry or ecology, but to integrate all functions.

Objective

The management has to obtain healthy and stable forest systems that produce wood with a minimum human intervention. The products to obtain, other than wood, are fauna habitats, biodiversity, recreational, aesthetics, and water management. The human action has the object of accelerating natural processes, but not substitute them.

Silvicultural Models

Forest with irregular structure and mixed composition, in NE Germany.

ProSilva recommends to use the *uneven-aged forest* system, in which the ages, and consequently sizes, of trees in a forest are different. It has the advantage to offer a stable structure regarding natural disasters and plagues, and is very adequate for fauna habitat and biodiversity promotion. It provides a better soil protection, since there is a permanent tree cover.

McEvoy considers that in spite of being the most close to nature system, it is difficult to implement, and proposes to use the *high regular forest* model, in which all trees are of the same age/size, but recommends using a regeneration system with a generous cover, to avoid soil erosion, and prevent excessive light entrance, that would promote the growth of a potent understorey.

The Ecoforestry Institute, similarly to Pro Silva, recommends multi-aged and multi-species forests.

Recommended Practices

Proposed thinning frequency is about ten years, and intensity low, in order to limit the ingress of excessive light, that could promote too much understory, or the growing of epicormic shoots. It has to be directed to favor the trees that show good prospect for the future. The operations have to be done in a way that will avoid soil compaction or damage to the trees that will remain standing.

Native/introduced Species

When foresters plant trees, they may use either native or introduced species. Whenever foresters decide to use a species that is not native, they do it because they think that there are silvicultural advantages linked to this choice, be they the wood quality, ease of management, adaptation to the climatic conditions, shorter production delay, etc. It may be that there is information available about the behavior of this species in the habitat, or the forester is ready to make a trial.

From the ecological point of view, the introducing species is considered as a threat. The introduced species risks being invasive. Invasives displace local species, resulting in a reduction of biodiversity, a condition also to be expected if great extensions are forested using introduced species.

Professor McEvoy is very clear and strict: introduced species can not be used at all when working in a *close to nature forestry* system. Pro Silva makes some distinctions, based on species and conditions. Natural forest systems are to be preserved, but the enrichment with certain introduced species may be positive, depending on circumstances.

ProSilva Recommendations

- The natural vegetation systems for each region are an asset to maintain, and constitute an important basis for the silvicultural planning.
- Introduced species can, depending on circumstances, supplement natural species, and add economical revenue.
- The forest species that are introduced to a region are termed exotic.
- The introduction of exotic species should only be permitted after trials conducted from the qualitative and quantitative point of view.

Herbivore Fauna Management

The herbivore fauna, be it domesticated or wild, acts on tree seedlings and small trees. In *high regular forests*, the regeneration periods are chosen by the forester, and therefore it is possible to establish some control over fauna action, particularly domesticated fauna, avoiding grazing during regeneration. The un-even aged forest is continuously regenerating, and therefore it is difficult to make it compatible with grazing, and does not admit a high density of wild herbivore fauna. The pressure of herbivore fauna, mainly cervids, in some European forests, has reached an intensity that is threatening the practice of *close to nature forestry*.

Economic Aspects

Forestry's economic profitability has progressively diminished in developed countries, beginning at the end of the 20th century. This has been the result of lower lumber prices and higher operating costs. Because it requires less human intervention, the *close to nature forestry* has lower labor costs. Also, it encourages the evolution of forests toward higher ecologic and landscape value structures. This is in demand by society, and the payment for ecosystem services is being considered.

Close to Nature Forestry/Sustainable Forestry

Close to nature forestry is a sustainable forestry, but the reverse does not apply. These are some of the distinctions between both systems.

- Sustainable forestry tries to preserve forests for the future, close to nature forestry aims to improve them.
- Sustainable forestry seeks a balance between competing interests, close to nature forestry integrates mutually interacting interests.
- Sustainable forestry aims to restrain forestry practice in order to preserve ecological values, close to nature forestry uses ecological principles to promote forestry results.

High Conservation Value Forest

High conservation value forest (HCVF) is a Forest Stewardship Council (FSC) forest management designation used to describe those forests who meet criteria defined by the FSC Principles and Criteria of Forest Stewardship.

Specifically, high conservation value forests are those that possess one or more of the following attributes:

1. forest areas containing globally, regionally or nationally significant: concentrations of biodiversity values (e.g. endemism, endangered species, refugia); and/or large landscape-level forests, contained within, or containing the management unit, where viable populations of most if not all naturally occurring species exist in natural patterns of distribution and abundance
2. forest areas that are in or contain rare, threatened or endangered ecosystems

3. forest areas that provide basic services of nature in critical situations (e.g. watershed protection, erosion control)

4. forest areas fundamental to meeting basic needs of local communities (e.g. subsistence, health) and/or critical to local communities' traditional cultural identity (areas of cultural, ecological, economic or religious significance identified in cooperation with such local communities).

History

The arose relatively recently in connection with the development of standards for the certification of forest management. The term was developed by the Forest Stewardship Council, A.C. (FSC), which is an international accreditation association1 incorporated in 1995. The FSC first started considering the term *high conservation values* in 1996 and formally included the term *high conservation value forests* in the FSC Principles and Criteria of Forest Stewardship in January 1999.

A number of terms have been used somewhat interchangeably with the term HCVF. To a certain extent this arises from translation between languages and to a certain extent it arises from varying approaches at interpretation. The Spanish version uses the terms *bosques con alto valor de conservación* (forests with high value of conservation) and *bosques con alto valor para la conservación* (forests with high value for conservation).

As noted, the term HCVF arose following consideration of the term *high conservation values*; this usage led to the concept of *forests having high conservation values* and *forests with high conservation values*. Indeed, the term *high conservation value* is sometimes given its own acronym HCV; however, the FSC refers to an HCV as "attributes".

Examples

Examples of the large landscape globally important forests include the Canada boreal forest and the remaining Amazon rainforest.

Examples of forest types which embrace rare or endangered species include the Dry deciduous forests of Madagascar the two Monterey Cypress of central coastal California and the Maritime Coast Range Ponderosa Pine forests of coastal California.

HCVF Toolkits

Following the inclusion of the term in the FSC P&C in 1999, the FSC formed an Advisory Panel for the Implementation of high conservation value forests and the precautionary principle. This panel produced a number of consultation documents on how to identify HCVF, including indicators and verifiers. However, this consultation process was not finalized and FSC has yet to issue any specific guidance regarding HCVF1.

In lieu of formal guidance from FSC, there has been a proliferation of discussion on HCVF, led mainly by non-governmental organisations (NGOs). In particular, the need for more precise and practical guidance led to a 2001 initiative by the UK-based consulting company, Proforest Ltd. This initiative led to the production of a three-part High Conservation Value Toolkit (the 'Global Toolkit'). The Global Toolkit offered a revised definition of HCVF as follows:

A High Conservation Value Forest is the area of forest required to maintain or enhance a High Conservation Value.

This definition provides a new approach to the zonation of HCVF as well as introduces the concept of 'High Conservation Value'. The inclusion of the term 'area of forest' provide clarity that there may be instances where an HCVF zone might be restricted to part of a forest, while the FSC definition implied that the presence of one or more HCV attribute would render the whole forest as to be a 'high conservation value forest'. Indeed, despite the provision for partial zonation, the Global Toolkit makes it clear that for the purposes of certification, any forest management unit containing even one small HCV is classified as a High Conservation Value Forest, which brings into play extra requirements for consultation, maintenance or enhancement, and annual monitoring. However, the additional burden is determined by the nature of the HCV and will affect only those parts of the forest, and those aspects of management, that are required for the maintenance or enhancement of the HCV (this procedure is explained in more detail in the section on Adaptive Management, below).

There are six recognized forms of High conservation values forests:

> HCV1. Forest areas containing globally, regionally or nationally significant concentrations of biodiversity values (e.g. endemism, endangered species, refugia).
>
> HCV2. Forest areas containing globally, regionally or nationally significant large landscape level forests, contained within, or containing the management unit, where viable populations of most if not all naturally occurring species exist in natural patterns of distribution and abundance.
>
> HCV3. Forest areas that are in or contain rare, threatened or endangered ecosystems.
>
> HCV4. Forest areas that provide basic services of nature in critical situations (e.g. watershed protection, erosion control).
>
> HCV5. Forest areas fundamental to meeting basic needs of local communities (e.g. subsistence, health).
>
> HCV6. Forest areas critical to local communities' traditional cultural identity (areas of cultural, ecological, economic or religious significance identified in co-operation with such local communities).

The Global Toolkit has not been explicitly endorsed by the FSC but the revised ordering system appears to have gained widespread appeal and there is a recommendation for its use in FSC reports found in a footnote to an FSC reporting standard. The Global Toolkit was intentionally designed for a wide range of users beyond the FSC scheme. Indeed, in December 2005, the popularity of the HCV approach led to the formation of the HCV Resource Network which includes a broad platform of NGOs as well as the World Bank and the International Tropical Timber Organization (ITTO).

High Conservation Value Areas

The concept of a High conservation value forest has now been generalized and used by other certification schemes like the Roundtable on Sustainable Palm Oil (RSPO), the Better Cotton Initiative, the Sustainable Agriculture Network (SAN) and Bonsucro, among others. Consumer goods companies (mainly members of the Consumer Goods Forum) have also made independent commitments to protecting High Conservation Values in their supply chains.

The analagous six High Conservation Values are:

- HCV 1: Concentrations of biological diversity including endemic species, and rare, threatened or endangered species, that are significant at global, regional or national levels.
- HCV 2: Intact forest landscapes and large landscape-level ecosystems and ecosystem mosaics that are significant at global, regional or national levels, and that contain viable populations of the great majority of the naturally occurring species in natural patterns of distribution and abundance.
- HCV 3: Rare, threatened, or endangered ecosystems, habitats or refugia.
- HCV 4: Basic ecosystem services in critical situations, including protection of water catchments and control of erosion of vulnerable soils and slopes.
- HCV 5: Sites and resources fundamental for satisfying the basic necessities of local communities or indigenous peoples (for livelihoods, health, nutrition, water, etc...), identified through engagement with these communities or indigenous peoples.
- HCV 6: Sites, resources, habitats and landscapes of global or national cultural, archaeological or historical significance, and/or of critical cultural, ecological, economic or religious/sacred importance for the traditional cultures of local communities or indigenous peoples, identified through engagement with these local communities or indigenous peoples..

References

- FAO (2015). Global Forest Resources Assessment 2015 How are the world's forests changing? (PDF). FAO. p. 7. ISBN 978-92-5-108821-0.

- Don, Gilmour (2016). FAO Forestry Paper 176: Forty years of community-based forestry: A review of its extent and effectiveness (PDF). FAO. ISBN 978-92-5-109095-4. Retrieved 5 April 2016.
- Barnard, Gerald Foley, Geoffrey (1984). Farm and community forestry. London: Earthscan, International Institute for Environment and Development. ISBN 0-905347-53-6.
- The Montréal Process. "Criteria and Indicators for the Conservation and Sustainable Management of Temperate and Boreal Forests" (PDF). www.montrealprocess.org. Retrieved 3 March 2016.
- "Report of the MCPFE Open-Ended Ad-Hoc Working Group on "Sustainability Criteria" For Forest Biomass Production, Including Bioenergy" (PDF). MCPFE Working Group. Retrieved 2 March 2016.
- The Montréal Process. "Montréal Process Criteria and Indicators" (PDF). www.montrealprocess.org. Retrieved 3 March 2016.
- Mushove, Patrick. "Stakeholder analysis in a community-based forest and wildlife resources management project in northern Mozambique.". Retrieved 24 September 2011.
- . "Indonesia Community-Based Forest Management". The Center for People and Forests, Bangkok, Thailand. Retrieved 24 September 2011.
- Food and Agriculture Organization of the United Nations. "Forestry Paper 64: Tree Growing by Rural People, 1985". Retrieved 24 September 2011.

3

Techniques and Approaches of Forest Conservation

Techniques and approaches are an important component of any field of study. The techniques mentioned in this chapter are afforestation, silviculture, reforestation, selection cutting and forest protection. Afforestation is the process of growing a forest in any area where there was no tree cover earlier. The following chapter elucidates the various techniques and approaches of forest conservation.

Afforestation

An afforestation project in Rand Wood, Lincolnshire, England

Afforestation is the establishment of a forest or stand of trees in an area where there was no forest. Reforestation is the reestablishment of forest cover, either naturally (by natural seeding, coppice, or root suckers) or artificially (by direct seeding or planting). Forestation is the establishment of forest growth on areas that either had forest or lacked it. So reforestation and afforestation are both forestation. Many governments and non-governmental organizations directly engage in programs of *afforestation* to create forests, increase carbon capture and sequestration, and help to anthropogenically improve biodiversity. (In the UK, afforestation may mean converting the legal status of some land to "royal forest".) Special tools, e.g. tree planting bar, are used to make planting of trees easier and faster.

Biological Process

Gap dynamics refers to the pattern of plant growth that occurs following the creation of a forest gap, a local area of natural disturbance that results in an opening in the canopy of a forest. Gap dynamics are a typical characteristic of both temperate and tropical forests and have a wide variety of causes and effects on forest life.

In Areas of Degraded Soil

In some places, forests need help to reestablish themselves because of environmental factors. For example, in arid zones, once forest cover is destroyed, the land may dry and become inhospitable to new tree growth. Other factors include overgrazing by livestock, especially animals such as goats, cows, and over-harvesting of forest resources. Together these may lead to desertification and the loss of topsoil; without soil, forests cannot grow until the long process of soil creation has been completed - if erosion allows this. In some tropical areas, forest cover removal may result in a duricrust or duripan that effectively seal off the soil to water penetration and root growth. In many areas, reforestation is impossible because people are using the land. In other areas, mechanical breaking up of duripans or duricrusts is necessary, careful and continued watering may be essential, and special protection, such as fencing, may be needed.

Countries and Regions

Afforested botanical garden in Hattori Ryokuchi Park, Japan.

Brazil

Due to the extensive and ongoing Amazon deforestation of the last few decades,)

China

China has deforested most of its historically wooded areas. China reached the point where timber yields declined far below historic levels, due to over-harvesting of trees beyond sustainable yield. Although it has set official goals for reforestation, these goals are set over an 80-years time horizon and have not been significantly met by 2008. China is trying to correct these problems by projects as the Green Wall of China, which aims to replant a great deal of forests and halt the expansion of the Gobi desert. A law promulgated in 1981 requires that every school student over the age of 11 plants at least one tree per year. As a result, China currently has the highest afforestation rate of any country or region in the world, with 47,000 square kilometers of afforestation in 2008. However, the forest area per capita is still far lower than the international average. There has also been considerable criticism regarding the effectiveness of planting so many trees especially in regions where they never grew prior. Studies reveal that the water table of those areas is becoming deeper indicating significant water loss.

India

Afforestation in South India

India has witnessed a minor increase in the percentage of the land area under forest cover from 1950 to 2006. In 1950 around 40.48 million hectares was covered by forest. In 1980 it increased to 67.47 million hectares and in 2006 it was found to be 69 million hectares. 23% of Indi a is covered by forest. The forests of India are grouped into 5 major categories and 16 types based on biophysical criteria. 38% of forest is categorised as subtropical dry deciduous and 30% as tropical moist deciduous plus other smaller groups. It is taken care that only local species are planted in an area. Trees bearing fruits are preferred wherever possible due to their function as a food source.

Hong Kong

Since the founding of the crown colony in the 19th century, afforestation has taken place to prevent soil erosion in the catchment areas of the reservoirs that were built. During the Japanese occupation in the Second World War, the countryside was deforested as the remaining population required fuel to survive. Most of the trees were cut down and extensive reafforestation was carried out after the war. Trees that were planted are mostly non-native species, such as: Pinus massoniana, Acacia confusa (Formosan acacia), Lophostemon confertus and the Paper Bark Tree.

Burkina-faso

Desertification is increasing along the Sahel, the strip of land between Africa's fertile tropics and the Sahara Desert. After a crippling famine in the 1970s caused by over-grazing and deforestation, a local community approach has been pioneered by Yacouba Sawadogo, a peasant farmer. By replanting trees and crops together in holes filled with compost, whole villages have been able to move back to areas considered uninhabitable.

Iran

Iran is considered a low forest cover region of the world with present cover approximating seven percent of the land area. This is a value reduced by an estimated six million hectares of virgin forest, which includes oak, almond and pistachio. Due to soil substrates, it is difficult to achieve afforestation on a large scale compared to other temperate areas endowed with more fertile and less rocky and arid soil conditions. Consequently, most of the afforestation is conducted with non-native species, leading to habitat destruction for native flora and fauna, and resulting in an accelerated loss of biodiversity.

JNF trees in the Negev Desert. Man-made dunes (here a liman) help keep in rainwater, creating an oasis.

Japan

Tree-planting is an ancient Jewish tradition, mentioned in the Talmud as being more important than greeting the Messiah. With over 240 million planted trees, Israel is one of only two countries that entered the 21st century with a net gain in the number of trees, due to massive afforestation efforts. Israeli forests are the product of a major afforestation campaign by the Jewish National Fund (JNF).

Critics argue that many JNF lands inside the West Bank were illegally confiscated from Palestinian refugees, and that the JNF furthermore should not be involved with lands in the West Bank. Shaul Ephraim Cohen has claimed that trees have been planted to restrict Bedouin herding. Susan Nathan wrote that forests were planted on the site of abandoned Arab villages after the 1948 war.

Since 2009, the JNF has provided the Palestinian Authority with 3,000 tree seedlings for a forested area being developed on the edge of the new city of Rawabi, north of Ramallah.

North Africa

In North Africa, the Sahara Forest Project coupled with the Seawater greenhouse has been proposed. Some projects have also been launched in countries as Senegal to revert desertification. As of 2010, African leaders are discussing the combining of national resources to increase effectiveness. In addition, other projects as the Keita Project in Niger have been launched in the past, and have been able to locally revert damage done by desertification.

Europe

Europe has deforested the majority of its historical forests. The European Union (EU) has paid farmers for afforestation since 1990, offering grants to turn farmland back into forest and payments for the management of forest. Between 1993 and 1997, EU afforestation policies made possible the re-forestation of over 5,000 square kilometres of land. A second program, running between 2000 and 2006, afforested more than 1000 square kilometres of land (precise statistics not yet available). A third such program began in 2007. Europe's forests are growing by 0.8 million ha a year thanks to these programmes.

In Poland, the National Program of Afforestation was introduced by the government after World War II, when area of forests shrank to 20% of country's territory. Consequently, forested areas of Poland grew year by year, and on December 31, 2006, forests covered 29% of the country. It is planned that by 2050, forests will cover 33% of Poland.

According to FAO statistics, Spain had the third fastest afforestation rate in Europe in the 1990-2005 period, after Iceland and Ireland. In those years, a total of 44,360 square kilometers were afforested, and the total forest cover rose from 13,5 to 17,9 million hectares. In 1990, forests covered 26,6% of the Spanish territory. As of 2007, that figure had risen to 36,6%. Spain today has the fifth largest forest area in the European Union.

In January 2013 the UK government set a target of 12% woodland cover in England by 2060, up from the then 10%. Government-backed initiatives such as the Woodland Carbon Code are intended to support this objective by encouraging corporations and landowners to create new woodland to offset their carbon emissions.

Alpine and Subalpine regions have undergone a lot of deforestation and then forestation in the last 300 years. Out of this has emerged much practical experience. One example is the Rotten group. whch is a method to bring in stable age mixed tree communities,

Australia

In Adelaide, South Australia (a city of 1.3 million) Premier Mike Rann (2002 to 2011) launched an urban forest initiative in 2003 to plant 3 million native trees and shrubs by 2014 on 300 project sites across the metro area. The projects range from large habitat restoration projects to local biodiversity projects. Thousands of Adelaide citizens have participated in community planting days. Sites include parks, reserves, transport corridors, schools, water courses and coastline. Only trees native to the local area are planted to ensure genetic integrity. Premier Rann said the project aimed to beautify and cool the city and make it more liveable; improve air and water quality and reduce Adelaide's greenhouse gas emissions by 600,000 tonnes of Co2 a year. He said it was also about creating and conserving habitat for wildlife and preventing species loss.

United States

The United States is roughly one-third covered in forest and woodland. Nevertheless, areas in the US were subject to significant tree planting. In the 1800s people moving westward encountered the Great Plains; land with fertile soil, a growing population and a demand for timber but with few trees to supply it. So tree planting was encouraged along homesteads. Arbor Day was founded in 1872 by Julius Sterling Morton in Nebraska City, Nebraska. By the 1930s the environmental disaster, the Dust Bowl signified a reason for significant tree cover. Public work's programs under the New Deal saw the planting of 18,000 miles of windbreaks stretching from North Dakota to Texas to fight soil erosion.

At their summit in Copenhagen in 2009, organised by the UK based The Climate Group, leaders of sub-national governments - States, Regions and Provinces - unanimously supported a recommendation by Premier Rann to plant 1 billion trees across their varied jurisdictions. The initiative was strongly supported by leaders present including Quebec Premier Jean Charest, California Governor Arnold Schwarzenegger and Scottish First Minister Alex Salmond. At a subsequent meeting in Rio de Janeiro in June 2012, The Climate Group announced that it had already received commitments by member governments to plant more than 500 million trees.

Silviculture

Silviculture is the practice of controlling the establishment, growth, composition, health, and quality of forests to meet diverse needs and values.

The name comes from the Latin *silvi-* (forest) + culture (as in growing). The study of forests and woods is termed silvology. Silviculture also focuses on making sure that the treatment(s) of forest stands are used to preserve and to better their productivity.

Generally, silviculture is the science and art of growing and tending forest crops, based on a knowledge of silvics, i.e., the study of the life history and general characteristics of forest trees and stands, with particular reference to locality factors. More particularly, silviculture is the theory and practice of controlling the establishment, composition, constitution, and growth of forests. No matter how forestry as a science is constituted, the kernel of the business of forestry is silviculture, as it includes direct action in the forest, and in it all economic objectives and technical considerations ultimately converge. The kernel of silviculture is regeneration.

Suggestions for how best to go about the job, presented by Jeglum et al. (2003), though aimed primarily at the boreal forest in Ontario, merit wider consideration. The 110-page publication describes Best Management Practices, first by general principles, then by sensitive sites. Illustrations are plentiful and are well chosen to complement this excellent text.

To some the distinction between forestry and silviculture is that silviculture is applied at the stand level and forestry is broader. For example, John D. Matthews says "complete regimes for regenerating, tending, and harvesting forests" are called "silvicultural systems".

So adaptive management is required for silviculture, whereas forestry can be natural, conserved land without a stand level treatment being applied. A common taxonomy divides silviculture into regenerating, tending and harvesting techniques.

An employee of the White Mountain Apache Tribe forestry department uses a hoedad to dig a hole for a ponderosa pine seedling in the shade of a burned tree stump, which protects the seedling from windy conditions that dry the trees

Silvicultural Systems

The origin of forestry in German speaking Europe has defined silvicultural systems broadly as Hochwald - High forest, Mittlewald - Coppice with standards (and Compound coppice), Short rotation coppice and Niederwald - Coppice. There are other systems too. These varied silvicultural systems include several harvesting methods, which are often wrongly said to be a silvicultural systems, but may also be called as rejuvenating or regenerating method depending on the purpose.

Hochwald is further subdivided in German into: Altersklassenwald - Age class forest which includes:

- *even aged forestry*
    - o - Kahlschlag - Clear cutting
    - o - Schirmschlag - Shelterwood cutting
    - o - Seed-tree method
- *un-even aged forestry*
    - o - Femelschlag (Femelschlagverfahren) - The Femel selection cutting (Group selection cutting),
    - o - Saumschlag (Saumfemelschlagbetrieb) - Strip selection cutting (strip-and-group felling system)
    - o - Schirmkeilschlag - Shelterwood wedge cutting
    - o - Mischformen - mixed form regeneration methods.

and Dauerwald - Continuous cover forestry, in which are found:

- *un-even aged forestry*
    - o - Plenterwald - Selection forest
    - o - Zielstärkennutzung - Target diameter harvesting.

From these names the impression is given that these are neatly defined systems, but in practice there are a variety of variations within these harvesting methods according to local ecology and site conditions. While an individual location can be identified where it could be said to be the archetypal form of harvesting technique (they all originated somewhere with a particular forester, and have been described in the scientific literature), and broad generalizations can be made these are merely rules of thumb rather than strict blue prints on how techniques might be applied. This misunderstanding has meant that many older English textbooks did not capture the true complexity of silviculture as practiced where it originated in Mitteleuropa.

This silviculture was culturally predicated on wood production in temperate and boreal climates and did not deal with tropical forestry. The misapplication of this philosophy to those tropical forests has been problematic. There is also an alternative silvicultural tradition which developed in Japan and thus created a different biocultural landscape called satoyama.

After harvesting comes regeneration, which may be split into Natural and Artificial. And tending which includes release treatments, pruning, thinning and intermediate treatments. As it can be imagined it maybe that any of these 3 phases (harvesting - regeneration - tending) may happen at the same time, within a stand, depending on the goal for that particular stand.

Regeneration

Regeneration is basic to the continuation of forested, as well as to the afforestation of treeless land. Regeneration can take place through self-sown seed ("natural regeneration"), by artificially sown seed, or by planted seedlings. In whichever case, the performance of regeneration depends on its growth potential and the degree to which its environment allows the potential to be expressed. Seed, of course, is needed for all regeneration modes, both for natural or artificial sowing and for raising planting stock in a nursery.

Natural regeneration is:

"Human-assisted natural regeneration" means establishment of a forest age class from natural seeding or sprouting in an area after harvesting in that area through selection cutting, shelter (or seed-tree) harvest, soil preparation, or restricting the size of a clear-cut stand to secure natural regeneration from the surrounding trees.

The process of natural regeneration involves the renewal of forests by means of self-sown seeds, root suckers, or coppicing. In natural forests, conifers rely almost entirely on regeneration through seed. Most of the broadleaves, however, are able to regenerate by the means of emergence of shoots from stumps (coppice) and broken stems.

Seedbed Requirements

Any seed, self-sown or artificially applied, requires a seedbed suitable for securing germination.

In order to germinate, a seed requires suitable conditions of temperature, moisture, and aeration. For seeds of many species, light is also necessary, and facilitates the germination of seeds in other species, but spruces are not exacting in their light requirements, and will germinate without light. White spruce seed germinated at 35 °F (1.7 °C) and 40 °F (4.4 °C) after continuous stratification for one year or longer and developed radicles < 6 cm long in the cold room. When exposed to light, those germinants developed chlorophyll and were normally phototropic with continued elongation.

For survival in the short and medium terms, a germinant needs: a continuing supply of moisture; freedom from lethal temperature; enough light to generate sufficient photosynthate to support respiration and growth, but not enough to generate lethal stress in the seedling; freedom from browsers, tramplers, and pathogens; and a stable root system. Shade is very important to the survival of young seedlings. In the longer term, there must be an adequate supply of essential nutrients and an absence of smothering.

In undisturbed forest, decayed windfallen stemwood provides the most favourable seedbed for germination and survival, moisture supply being dependable, and the elevation of seedlings somewhat above the general level of the forest floor reduces the danger of smothering by leaves and snow-pressed minor vegetation; nor is such a microsite likely to be subject to flooding. Advantages conferred by those microsites include: more light, higher temperatures in the rooting zone, and better mycorrhizal development. In a survey in the Porcupine Hills, Manitoba, 90% of all spruce seedlings were rooted in rotten wood.

Mineral soil seedbeds are more receptive than the undisturbed forest floor, and are generally moister and more readily rewetted than the organic forest floor. However, exposed mineral soil, much more so than organic-surfaced soil, is subject to frost heaving and shrinkage during drought. The forces generated in soil by frost or drought are quite enough to break roots.

The range of microsites occurring on the forest floor can be broadened, and their frequency and distribution influenced by site preparation. Each microsite has its own microclimate. Microclimates near the ground are better characterized by vapour pressure deficit and net incident radiation, rather than the standard measurements of air temperature, precipitation, and wind pattern.

Aspect is an important component of microclimate, especially in relation to temperature and moisture regimes. Germination and seedling establishment of Engelmann spruce were much better on north than on south aspect seedbeds in the Fraser Experimental Forest, Colorado; the ratios of seeds to 5-year-old seedlings were determined as 32:1, 76:1, and 72:1 on north aspect bladed-shaded, bladed-unshaded, and undisturbed-shaded seedbeds, respectively. Clearcut openings of 1.2 ha to 2.0 ha adjacent to an adequate seed source, and not more than 6 tree heights wide, could be expected to secure acceptable regeneration (4,900, 5-year-old trees/ha), whereas on undisturbed-unshaded north aspects, and on all seedbed treatments tested on south aspects, seed to seedling ratios were so high that the restocking of any clearcut opening would be questionable.

At least 7 variable factors may influence seed germination: (1) seed characteristics, (2) light, (3) oxygen, (4) soil reaction (ph), (5) temperature, (6) moisture, and (7) seed enemies. Moisture and temperature are the most influential, and both are affected by exposure. The difficulty of securing natural regeneration of Norway spruce and Scots pine in northern Europe led to the adoption of various forms of reproduction cuttings that provided partial shade or protection to seedlings from hot sun and wind. The main objective of echeloned strips or border-cuttings with northeast exposure was to protect regeneration from overheating, and was originated in Germany and deployed successfully by Alarik (1925) and others in Sweden. On south and west exposures, direct insolation and heat reflected from tree trunks often result in temperatures lethal to young seedlings, as well as desiccation of the surface soil, which inhibits germination. The sun is less injurious on eastern exposures because of the lower temperature in the early morning, related to higher humidity and presence of dew.

Baldwin (1933), after noting that summer temperatures in North America are often higher than those in places where border-cuttings have been found useful, reported the results of a survey of regeneration in a stand of red spruce plus scattered white spruce that had been isolated by clearcutting on all sides, so furnishing an opportunity for observing regeneration on different exposures in this old-field stand at Dummer, New Hampshire. The regeneration included a surprisingly large number of balsam fir seedlings from the 5% stand component of that species. The maximum density of spruce regeneration, determined 4 rods (20 m) inside from the edge of the stand on a North 20°E exposure, was 600,000/ha, with almost 100,000 balsam fir seedlings.

A prepared seedbed remains receptive for a relatively short period, seldom as long as 5 years, sometimes as short as 3 years. Seedbed receptivity on moist, fertile sites decreases with particular rapidity, and especially on such sites, seedbed preparation should be scheduled to take advantage of good seed years. In poor seed years, site preparation can be carried out on mesic and drier sites with more chance of success, because of the generally longer receptivity of seedbeds there than those on moister sites. Although an indifferent seed year can suffice if seed distribution is good and environmental conditions favourable to seedling germination and survival, small amounts of seed are particularly vulnerable to depredation by small mammals. Considerable flexibility is possible in timing site preparation to coincide with cone crops. Treatment can be apllied either before any logging takes place, between partial cuts, or after logging. In cut and leave strips, seedbed preparation can be carried out as a single operation, pre-scarifying the leave strips, post-scarifying the cut strips.

Broadcast burning is not recommended as a method of preparing sites for natural regeneration, as it rarely exposes enough mineral soil to be sufficiently receptive, and the charred organic surfaces are a poor seedbed for spruce. A charred surface may get too hot for good germination and may delay germination until fall, with subsequent overwinter mortality of unhardened seedlings. Piling and burning of logging slash, however, can leave suitable exposures of mineral soil.

Season of Planting

Artificial Regeneration

With a view to reducing the time needed to produce planting stock, experiments were carried out with white spruce and three other coniferous species from Wisconsin seed in the longer, frost-free growing season in Florida, 125 vs. 265 days in central Wisconsin and northern Florida, respectively. As the species studied are adapted to long photoperiods, extended daylengths of 20 hours were applied in Florida. Other seedlings were grown under extended daylength in Wisconsin and with natural daylength in both areas. After two growing seasons, white spruce under long days in Florida were about the same as those in Wisconsin, but twice as tall as plants under natural Wisconsin photoperiods. Under natural days in Florida, with the short local photoperiod, white spruce was severely dwarfed and had a low rate of survival. Black spruce responded similarly. After two growing seasons, long day plants of all 4 species in Florida were well balanced, with good development of both roots and shoots, equaling or exceeding the minimum standards for 2+1 and 2+2 outplanting stock of Lake States species. Their survival when lifted in February and outplanted in Wisconsin equalled that of 2+2 Wisconsin-grown transplants. Artificial extension of the photoperiod in the northern Lake States greatly increased height increment of white and black spruces in the second growing season.

Optimum conditions for seedling growth have been determined for the production of containerized planting stock. Alternating day/night temperatures have been found more suitable than a constant temperature; at 400 lumens/m^2 light regime, a 28 °C/20 °C day/night temperatures have been recommended for white spruce. However, temperature optima are not necessarily the same at different ages and sizes. Tinus (1984) investigated the effects of combinations of day and night temperature on height, caliper, and dry weight of 4 seed sources of Engelmann spruce. The 4 seed sources appeared to have very similar temperature requirements, with night optima about the same of slightly lower than daylight optima.

Tree provenance is important in artificial regeneration. Good provenance takes into account suitable tree genetics and a good environmental fit for planted / seeded trees in a forest stand. The wrong genotype can lead to failed regeneration, or poor trees that are prone to pathogens and undesired outcomes.

Artificial regeneration has been a more common method involving planting because it is more dependable than natural regeneration. Planting can involve using seedlings (from a nursery), (un)rooted cuttings, or seeds.

Whichever method is chosen it can be assisted by tending techniques also known as intermediate stand treatments.

The fundamental genetic consideration in artificial regeneration is that seed and planting stock must be adapted to the planting environment. Most commonly, the method of managing seed and stock deployment is through a system of defined seed zones, within which seed and stock can be moved without risk of climatic maladaptation. Ontario adopted a seed zone system in the 1970s based on Hills' (1952) site regions and provincial resource district boundaries, but Ontario's seed zones are now based on homogeneous climatic regions developed with the Ontario Climate Model. The regulations stipulate that source-identified seedlots may be either a general collection, when only the seed zone of origin is known, or a stand collection from a specific latitude and longitude. The movement of general-collection seed and stock across seed zone boundaries is prohibited, but the use of stand-collection seed and stock in another seed zone is acceptable when the Ontario Climate Model shows that the planting site and place of seed origin are climatically similar. The 12 seed zones for white spruce in Quebec are based mainly on ecological regions, with a few modifications for administrative convenience.

Seed quality varies with source. Seed orchards produce seed of the highest quality, then, in order of decreasing seed quality produced, seed production areas and seed collection areas follow, with controlled general collections and uncontrolled general collections producing the least characterized seed.

Seeds

Dewinging, Extraction

When seed is first separated from cones it is mixed with foreign matter, often 2 to 5 times the volume of the seed. The more or less firmly attached membranous wings on the seed must be detached before it is cleaned of foreign matter. The testa must not incur damage during the dewinging process. Two methods have been used, dry and wet. Dry seed may be rubbed gently through a sieve that has a mesh through which only seed without wings can pass. Large quantities of seed can be processed in dewinging machines, which use cylinders of heavy wire mesh and rapidly revolving stiff brushes within to remove the wings. In the wet process, seed with wings attached are spread out 10 cm to 15 cm deep on a tight floor and slightly moistened throughout; light leather flails are used to free seed from the wings. Wang (1973) described a unique wet dewinging procedure using a cement mixer, used at the Petawawa tree seed processing facility. Wings of white and Norway spruce seed can be removed by dampening the seed slightly before it is run through a fanning mill for the last time. Any moistened seed must be dried before fermentation or moulding sets in.

Seed Viability

A fluorescein diacetate (FDA) biochemical viability test for several species of conifer seed, including white spruce, estimates the proportion of live seed (viability) in a seedlot, and hence the percentage germination of a seedlot. The accuracy of predicting percentage germination was within +/- 5 for most seedlots. White spruce seed can be tested for viability by an indirect method, such as the fluorescein diacetate (FDA) test or 'Ultra-sound'; or by the direct growth method of 'germination'. Samples of white spruce seed inspected by Toumey and Stevens (1928) varied in viability from 50% to 100%, but averaged 93%. Rafn (1915) reported 97% viability for white spruce seed.

Germinative Testing

The results of a germination test are commonly expressed as *germinative capacity* or a *germination percentage*, which is the percentage of seeds that germinate during a period of time, ending when germination is practically complete. During extraction and processing, white spruce seeds gradually lost moisture, and total germination increased. Mittal et al. (1987) reported that white spruce seed from Algonquin Park, Ontario, obtained the maximum rate (94% in 6 days) and 99% total germination in 21 days after 14-week pre-chilling. The pre-treatment of 1% sodium hypochlorite increased germinability.

Encouraged by Russian success in using ultrasonic waves to improve the germinative energy and percentage germination of seeds of agricultural crops, Timonin (1966) demonstrated benefits to white spruce germination after exposure of seeds to 1, 2, or 4 minutes of ultrasound generated by an M.S.E. ultrasonic disintegrator with a power consumption of 280 VA and power impact of 1.35 amperes. However, no seeds germinated after 6 minutes of exposure to ultrasound.

Seed Dormancy

Seed dormancy is a complex phenomenon and is not always consistent within species. Cold stratification of white spruce seed to break dormancy has been specified as a requirement, but Heit (1961) and Hellum (1968) regarded stratification as unnecessary. Cone handling and storage conditions affect dormancy in that cold, humid storage (5 °C, 75% to 95% relative humidity) of the cones prior to extraction seemingly eliminated dormancy by overcoming the need to stratify. Periods of cold, damp weather during the period of cone storage might provide natural cold (stratification) treatment. Once dormancy was removed in cone storage, subsequent kiln-drying and seed storage did not reactivate dormancy.

Haddon and Winston (1982) found a reduction in viability of stratified seeds after 2 years of storage and suggested that stress might have been caused by stratification, e.g., by changes in seed biochemistry, reduced embryo vigor, seed aging or actual damage to the embryo. They further questioned the quality of the 2-year-old seed even though high germination occurred in the samples that were not stratified.

Cold Stratification

Cold stratification is the term applied to the storing of seeds in (and, strictly, in layers with) a moist medium, often peat or sand, with a view to maintaining viability and overcoming dormancy. Cold stratification is the term applied to storage at near-freezing temperatures, even if no medium is used. A common method of cold stratification, is to soak seed in tap water for up to 24 h, superficially dry it, then store moist for some weeks or even months at temperatures just above freezing. Although Hellum (1968) found that cold stratification of an Alberta seed source led to irregular germination, with decreasing germination with increasing length of the stratification period, Hocking's (1972) paired test with stratified and nonstratified Alberta seed from several sources revealed no trends in response to stratification. Hocking suggested that seed maturity, handling, and storage needed to be controlled before the need for stratification could be determined. Later, Winston and Haddon (1981) found that the storage of white spruce cones for 4 weeks at 5 °C prior to extraction obviated the need for stratification.

Seed Ripeness

Seed maturity cannot be predicted accurately from cone flotation, cone moisture content, cone specific gravity; but the province of B.C. found embryo occupying 90% + of the corrosion cavity and megagametophyte being firm and whitish in colour are the best predictors for white spruce in B.C., and Quebec can forecast seed maturity some weeks in advance by monitoring seed development in relation to heat-sums and the phenological progression of the inflorescence of fireweed (*Epilobium angustifolium* L.), an associated plant species. Cone collection earlier than one week before seed maturity would reduce seed germination and viability during storage. Four stages of maturation were determined by monitoring carbohydrates, polyols, organic acids, respiration, and metabolic activity. White spruce seeds require a 6-week post-harvest ripening period in the cones to obtain maximum germinability, however, based on cumulative degree-days, seed from the same trees and stand showed 2-week cone storage was sufficient.

Forest tree Plantations

Plantation Establishment Criteria

Plantations may be considered successful when outplant performance satisfies certain criteria. The term "free growing" is applied in some jurisdictions. Ontario's "Free-to-Grow" (FTG) equivalent relates to a forest stand that meets a minimum stocking standard and height requirement, and is essentially free of competition from surrounding vegetation that might impede growth. The FTG concept was introduced with the advent of the Forest Management Agreement program in Ontario in 1980 and became applicable to all management units in 1986. Policy, procedures, and methodologies readily applicable by forest unit managers to assess the effectiveness of regeneration programs were still under development during the Class Environmental Assessment hearings.

In British Columbia, the Forest Practices Code (1995) governs performance criteria. To minimize the subjectivity of assessing deciduous competition as to whether or not a plantation is established, minimum specifications of number, health, height, and competition have been specified in British Columbia. However, minimum specifications are still subjectively set and may need to be fine-tuned in order to avoid unwarranted delay in according established status to a plantation. For example, a vigorous white spruce with a strong, multi-budded leading shoot and its crown fully exposed to light on 3 sides would not qualify as free-growing in the current British Columbia Code but would hardly warrant description as unestablished.

Competition

Competition arises when individual organisms are sufficiently close together to incur growth constraint through mutual modification of the local environment. Plants may compete for light, moisture and nutrients, but seldom for space *per se*. Vegetation management directs more of the site's resources into usable forest products, rather than just eliminating all competing plants. Ideally, site preparation ameliorates competition to levels that relieve the outplant of constraints severe enough to cause prolonged check.

The diversity of boreal and sub-boreal broadleaf-conifer mixed tree species stands, commonly referred to as the "mixedwoods", largely preclude the utility of generalizations and call for the development of management practices incorporating the greater inherent complexity of broadleaf-conifer mixtures, relative to single-species or mixed-species conifer forest. After harvesting or other disturbance, mixedwood stands commonly enter a prolonged period in which hardwoods overtop the coniferous component, subjecting them to intense competition in an understorey. It is well established that the regeneration and growth potential of understorey conifers in mixedwood stands is correlated to the density of competing hardwoods. To help apply "free-to-grow" regulations in British Columbia and Alberta, management guidelines based on distance-dependent relations within a limited radius of crop trees were developed, but Lieffers et al. (2002) found that free-growing stocking standards did not adequately characterize light competition between broadleaf and conifer components in boreal mixedwood stands, and further noted that adequate sampling using current approaches would be operationally prohibitive.

Many promising plantations have failed through lack of tending. Young crop trees are often ill-equipped to fight it out with competition resurgent following initial site preparation and planting.

Perhaps the most direct evaluation of the effect of competition on plantation establishment is provided by an effective herbicide treatment. The fact that herbicide treatment does not always produce positive results should not obscure the demonstrated potential of herbicides for significantly promoting plantation establishment. Factors that can vitiate the effectiveness of a herbicide treatment include: weather, especially temperature, prior to and during application; weather, especially wind, during application; weather, especially precipitation, in the 12 to 24 hours after application; vegetation characteristics, including species, size, shape, phenological stage, vigour, and distribution of weeds; crop characteristics, including species, phenology, and condition; the effects of other treatments, such as preliminary shearblading, burning or other prescribed or accidental site preparation; and the herbicide used, including dosage, formulation, carrier, spreader, and mode of application. There is a lot that can go wrong, but a herbicide treatment can be as good or better than any other method of site preparation.

Competition Indices

The study of competition dynamics requires both a measure of the competition level and a measure of crop response. Various competition indices have been developed, e.g., by Bella (1971) and Hegyi (1974) based on stem diameter, by Arney (1972), Ek and Monserud (1974), and Howard and Newton (1984) based on canopy development, and Daniels (1976), Wagner (1982), and Weiner (1984) with proximity-based models.

Studies generally considered tree response to competition in terms of absolute height or basal area, but Zedaker (1982) and Brand (1986) sought to quantify crop tree size and environmental influences by using relative growth measures.

Tending

Tending is the term applied to pre-harvest silvicultural treatment of forest crop trees at any stage after initial planting or seeding. The treatment can be of the crop itself (e.g., spacing, pruning, thinning, and improvement cutting) or of competing vegetation (e.g., weeding, cleaning).

Planting

How many trees per unit area (spacing) that should be planted is not an easily answered question. Establishment density targets or regeneration standards have commonly been based on traditional practice, with the implicit aim of getting the stand quickly to the free-to-grow stage. Money is wasted if more trees are planted than are needed to achieve desired stocking rates, and the chance to establish other plantations is proportionately diminished. Ingress (natural regeneration) on a site is difficult to predict and often becomes surprisingly evident only some years after planting has been carried out. Early stand development after harvesting or other disturbance undoubtedly varies greatly among sites, each of which has its own peculiar characteristics.

For all practical purposes, the total volume produced by a stand on a given site is constant and optimum for a wide range of density or stocking. It can be decreased, but not increased, by altering the amount of growing stock to levels outside this range. Initial density affects stand development in that close spacing leads to full site utilization more quickly than wider spacing. Economic operability can be advanced by wide spacing even if total production is less than in closely spaced stands.

Beyond the establishment stage, the relationship of average tree size and stand density is very important. Various density-management diagrams conceptualizing the density-driven stand dynamics have been developed. Smith and Brand's (1988) diagram has mean tree volume on the vertical axis and the number of trees/ha on the horizontal axis: a stand can either have many little trees or a few big ones. The self-thinning line shows the largest number of trees of a given size/ha that can be carried at any given time. However, Willcocks and Bell (1995) caution against using such diagrams unless specific knowledge of the stand trajectory is known.

In the Lake States, plantations have been made with the spacing between trees varying from 3 by 3 to 10 by 10 feet (0.9 m by 0.9 m to 3.0 m by 3.0 m). Kittredge recommended that no fewer than 600 established trees per acre (1483/ha) be present during the early life of a plantation. To insure this, at least 800 trees per acre (1077/ha) should be planted where 85% survival may be expected, and at least 1200/ac (2970/ha) if only half of them can be expected to live. This translates into recommended spacings of 5 by 5 to 8 by 8 feet (1.5 m by 1.5 m to 2.4 m by 2.4 m) for plantings of conifers, including white spruce in the Lake States.

Enrichment Planting

A strategy for enhancing natural forests' economic value is to increase their concentration of economically important, indigenous tree species by planting seeds or seedlings for future harvest, which can be accomplished with enrichment planting (EP). This means increasing the planting density (i.e., the numbers of plants per hectare) in an already growing forest stand."

Release Treatments

- Weeding: A process of getting rid of saplings' or seedlings' competition by mowing, application of herbicide, or other method of removal from the surroundings.
- Cleaning: Release of select saplings from competition by overtopping trees of a comparable age. The treatment favors trees of a desired species and stem quality.
- Liberation cutting: A treatment that releases tree seedling or saplings by removing older overtopping trees.

Spacing

Over-crowded regeneration tends to stagnate. The problem is aggravated in species that have little self-pruning ability, such as white spruce. Spacing is a thinning (of natural regeneration), in which all trees other than those selected for retention at fixed intervals are cut. The term juvenile spacing is used when most or all of the cut trees are unmerchantable. Spacing can be used to obtain any of a wide range of forest management objectives, but it is especially undertaken to reduce density and control stocking in young stands and prevent stagnation, and to shorten the rotation, i.e., to speed the production of trees of a given size. Volume growth of individual trees and the merchantable growth of stands are increased. The primary rationale for spacing is that thinning is the projected decline in maximum allowable cut. And since wood will be concentrated on fewer, larger, and more uniform stems, operating and milling costs will be minimized.

Methods for spacing may be: manual, using various tools, including power saws, brush saws, and clippers; mechanical, using choppersand mulchers; chemical; or combinations of several methods. One treatment has had notable success in spacing massively overstocked (<100 000 stems/ha) natural regeneration of spruce and fir in Maine. Fitted to helicopter, the Thru-Valve boom emits herbicide spray droplets 1000 µm to 2000 µm in diameter at very low pressure. Swaths 1.2 m wide and leave strips 2.4 m wide were obtained with "knife-edge" precision when the herbicide was applied by helicopter flying at a height of 21 m at a speed of 40–48 km/h. It seems likely that no other method could be as cost-effective.

Twenty years after spacing to 2.5 × 2.5 m, 30-year-old mixed stands of balsam fir and white spruce in the Green River watershed, New Brunswick, averaged 156.9 m^3/ha.

A spacing study of 3 conifers (white spruce, red pine and jack pine) was established at Moodie, Manitoba, on flat, sandy, nutritionally poor soils with a fresh moisture regime. Twenty years after planting, red pine had the largest average dbh, 15% greater than jack pine, while white spruce dbh was less than half that of the pines. Crown width showed a gradual increase with spacing for all 3 conifers. Results to date were suggesting optimum spacings between 1.8 m and 2.4 m for both pines; white spruce was not recommended for planting on such sites.

Comparable data are generated by espacement trials, in which trees are planted at a range of densities. Spacings of 1.25 m, 1.50 m, 1.75 m, 2.00 m, 2.50 m, and 3.00 m on 4 site classes were used in the 1922 trial at Petawawa, Ontario. In the first of 34 old field white spruce plantations used to investigate stand development in relation to spacing at Petawawa, Ontario, regular rows were planted at average spacings of from 4 × 4 to 7 × 7 feet (1.22 m × 1.22 m to 2.13 m × 2.13 m). Spacings up to 10 × 10 feet (3.05 m × 3.03 m) were subsequently included in the study. Yield tables based on 50 years of data showed:

a) Except for merchantable volumes at age 20 and site classes 50 and 60, closer spacings gave greater standing volumes at all ages than did wider spacings, the relative difference decreasing with age.

b) Merchantable volume as a proportion of total volume increases with age, and is greater at wider than at closer spacings.

c) Current annual volume increment culminates sooner at closer than at wider spacings.

A smaller espacement trial, begun in 1951 near Thunder Bay, Ontario, included white spruce at spacings of 1.8 m, 2.7 m, and 3.6 m. At the closest spacing, mortality had begun at 37 years, but not at the wider spacings.

The oldest interior spruce espacement trial in British Columbia was established in 1959 near Houston in the Prince Rupert Forest Region. Spacings of 1.2 m, 2.7 m, 3.7 m, and 4.9 m were used, and trees were measured 6, 12, 16, 26, and 30 years after planting. At wide espacements, trees developed larger diameters, crowns, and branches, but (at 30 years) basal area and total volume/ha were greatest in the closest espacement (Table 6.38). In more recent trials in the Prince George Region of British Columbia (Table 6.39) and in Manitoba, planting density of white spruce had no effect on growth after up to 16 growing seasons, even at spacings as low as 1.2 m. The slowness of juvenile growth and of crown closure delay the response to intra-competition. Initially, close spacing might even provide a positive nurse effect to offset any negative response to competition.

Thinning

Thinning is an operation that artificially reduces the number of trees growing in a stand with the aim of hastening the development of the remainder. The goal of thinning is to control the amount and distribution of available growing space. By altering stand density, foresters can influence the growth, quality, and health of residual trees. It also provides an opportunity to capture mortality and cull the commercially less desirable, usually smaller and malformed, trees. Unlike regeneration treatments, thinnings are not intended to establish a new tree crop or create permanent canopy openings.

Thinning greatly influences the ecology and micro-meteorology of the stand, lowering the inter-tree competition for water. The removal of any tree from a stand has repercussions on the remaining trees both above-ground and below. Silvicultural thinning is a powerful tool that can be used to influence stand development, stand stability, and the characteristics of the harvestable products.

When considering intensive conifer plantations designed for maximum production, it is essential to remember that tending and thinning regimes and wind and snow damage are intimately related.

Previous studies have demonstrated that repeated thinnings over the course of a forest rotation increase carbon stores relative to stands that are clear-cut on short rotations and that the carbon benefits differ according to thinning method (e.g., thinning from above versus below).

Precommercial Thinning

When natural regeneration or artificial seeding has resulted in dense, overstocked young stands, natural thinning will in most cases eventually reduce stocking to more silviculturally desirable levels. But by the time some trees reach merchantable size, others will be overmature and defective, and others will still be unmerchantable. The yield of merchantable wood can be greatly increased and the rotation shortened by precommercial thinning. Mechanical and chemical methods have been applied, but their costliness has militated against their ready adoption.

Pruning

Pruning, as a silvicultural practice, refers to the removal of the lower branches of the young trees (also giving the shape to the tree) so clear knot-free wood can subsequently grow over the branch stubs. Clear knot-free lumber has a higher value. Pruning has been extensively carried out in the Radiata pine plantations of New Zealand and Chile, however the development of Finger joint technology in the production of lumber and mouldings has led to many forestry companies reconsidering their pruning practices. "Brashing" is an alternative name for the same process. Pruning can be done to all trees, or more cost effectively to a limited number of trees. There are two types of pruning: natural or self-pruning and artificial pruning. Most cases of self-pruning happen when branches do not receive enough sunlight and die. Wind can also take part in natural pruning which can break branches. Artificial pruning is where people are paid to come and cut the branches. Or it can be natural, where trees are planted close enough that the effect is to cause self-pruning of low branches as energy is put into growing up for light reasons and not branchiness.

Stand Conversion

The term *stand conversion* refers to a change from one silvicultural system to another and includes *species conversion*, i.e., a change from one species (or set of species) to another. Such change can be effected intentionally by various silvicultural means, or incidentally by default e.g., when high-grading has removed the coniferous content from a mixedwood stand, which then becomes exclusively self-perpetuating aspen. In general, such sites as these are the most likely to be considered for conversion.

Growth and Yield

In discussing yields that might be expected from the Canadian spruce forests, Haddock (1961) noted that Wright's (1959) quotation of spruce yields in the British Isles of 220 cubic feet per acre (15.4 m^3/ha) per year and in Germany of 175 cubic feet per acre (12.25 m^3/ha) per year was misleading, at least if it was meant to imply that such yields might be approached in the Boreal Forest Region of Canada. Haddock thought that Wright's suggestion of 20 to 40 (average 30) cubic feet per acre (1.4 m^3/ha to 2.8 m^3/ha (average 2.1 m^3/ha) per year was more reasonable, but still somewhat optimistic.

The principal way forest resource managers influence growth and yield is to manipulate the mixture of species and number (density) and distribution (stocking) of individuals that form the canopy of the stand. Species composition of much of the boreal forest in North America already differs greatly from its pre-exploitation state. There is less spruce and more hardwoods in the second-growth forest than in the original forest; Hearnden et al. (1996) calculated that the spruce cover type had declined from 18% to only 4% of the total forested area in Ontario. Mixedwood occupies a greater proportion of Ontario's second-growth forest (41%) than in the original (36%), but its component of white spruce is certainly much diminished.

Growth performance is certainly influenced by site conditions and thus by the kind and degree of site preparation in relation to the nature of the site. It is important to avoid the assumption that site preparation of a particular designation will have a particular silvicultural outcome. Scarification, for instance, not only covers a wide range of operations that scarify, but also any given way of scarifying can have significantly different results depending on site conditions at the time of treatment. In point of fact, the term is commonly misapplied. *Scarification* is defined as "Loosening the top soil of open areas, or breaking up the forest floor, in preparation for regenerating by direct seeding or natural seedfall", but the term is often misapplied to practices that include scalping, screefing, and blading, which pare off low and surface vegetation, together with most off its roots to expose a weed-free surface, generally in preparation for sowing or planting thereon.

Thus, it is not surprising that literature can be used to support the view that the growth of seedlings on scarified sites is much superior to that of growth on similar sites that have not been scarified, while other evidence supports the contrary view that scarification can reduce growth. Detrimental results can be expected from scarification that impoverishes the rooting zone or exacerbates edaphic or climatic constraints.

Burning site preparation has enhanced spruce seedling growth, but it must be supposed that burning could be detrimental if the nutrient capital is significantly depleted.

An obvious factor greatly influencing regeneration is competition from other vegetation. In a pure stand of Norway spruce, for instance, Roussel (1948) found the following relationships:

Percent cover (%)	Vegetation Description
Below 1	No vegetation
1-3	Moss carpet with a few fir seedlings
4-10	Herbaceous plants appear
10-25	Bramble, herbs, fairly vigorous spruce seedlings
>25	Herbs, brambles very dense, vigorous, no moss

A factor of some importance in solar radiation–reproduction relationships is excess heating of the soil surface by radiation. This is especially important for seedlings, such as spruce, whose first leaves do not shade the base of the stem at the soil surface. Surface temperatures in sandy soils on occasion reach lethal temperatures of 50 °C to 60 °C.

Common Methods of Harvesting

Silvicultural regeneration methods combine both the harvest of the timber on the stand and re-establishment of the forest. The proper practice of sustainable forestry should mitigate the potential negative impacts, but all harvest methods will have some impacts on the land and residual stand. The practice of sustainable forestry limits the impacts such that the values of the forest are maintained in perpetuity. Silvicultural prescriptions are specific solutions to a specific set of circumstances and management objectives. Following are some common methods:

Clearcut Harvesting

Conventional clearcut harvesting is relatively simple: all trees on a cutblock are felled and bunched with bunches aligned to the skidding direction, and a skidder then drags the bunches to the closest log deck. Feller-buncher operators concentrate on the width of the felled swath, the number of trees in a bunch, and the alignment of the bunch. Providing a perimeter boundary is felled during daylight, night-shift operations can continue without the danger of trespassing beyond the block. Productivity of equipment is maximized because units can work independently of one another.

Clearcutting

An even-aged regeneration method that can employ either natural or artificial regeneration. It involves the complete removal of the forest stand at one time. Clearcutting can be biologically appropriate with species that typically regenerate from stand replacing fires or other major disturbances, such as Lodgepole Pine (*Pinus contorta*). Alternatively, clearcutting can change the dominating species on a stand with the introduction of non-native and invasive species as was shown at the Blodgett Experimental Forest near Georgetown California. Additionally, clearcutting can prolong slash decomposition, expose soil to erosion, impact visual appeal of a landscape and remove essential wildlife habitat. It is particularly useful in regeneration of tree species such as Douglas-fir (*Pseudotsuga menziesii*) which is shade intolerant. In addition, the general public's distaste for even-aged silviculture, particularly clearcutting, is likely to result in a greater role for uneven-aged management on public lands as well. Across Europe, and in parts of North America, even-aged, production-orientated and intensively managed plantations are beginning to be regarded in the same way as old industrial complexes: something to abolish or convert to something else.

Clearcutting will impact many site factors important in their effect on regeneration, including air and soil temperatures. Kubin and Kemppainen (1991), for instance, measured temperatures in northern Finland from 1974 through 1985 in 3 clear-felled areas and in 3 neighouring forest stands dominated by Norway spruce. Clear felling had no significant influence on air temperature at 2 m above the ground surface, but the daily air temperature maxima at 10 cm were greater in the clear-felled area than in the uncut forest, while the daily minima at 10 cm were lower. Night frosts were more common in the clear-felled area. Daily soil temperatures at 5 cm depth were 2 °C to 3 °C greater in the clear-felled area than in the uncut forest, and temperatures at depths of 50 cm and 100 cm were 3 °C to 5 °C greater. The differences between the clear-felled and uncut areas did not diminish during the 12 years following cutting.

Coppicing

A regeneration method which depends on the sprouting of cut trees. Most hardwoods, the coast redwood, and certain pines naturally sprout from stumps and can be managed through coppicing. Coppicing is generally used to produce fuelwood, pulpwood, and other products dependent on small trees. A close relative of coppicing is pollarding. Three systems of coppice woodland management are generally recognized: simple coppice, coppice with standards, and the coppice selection system.

- In Compound coppicing or coppicing with standards, some of the highest quality trees are retained for multiple rotations in order to obtain larger trees for different purposes.

Direct Seeding

Prochnau (1963), 4 years after sowing, found that 14% of viable white spruce seed sown on mineral soil had produced surviving seedlings, at a seed:seedling ratio of 7.1:1. With Engelmann spruce, Smith and Clark (1960) obtained average 7th year seed:seedling ratios of 21:1 on scarified seedbeds on dry sites, 38:1 on moist sites, and 111:1 on litter seedbeds.

Group Selection

The group selection method is an uneven-aged regeneration method that can be used when mid-tolerant species regeneration is desired. The group selection method can still result in residual stand damage in dense stands, however directional falling can minimize the damage. Additionally, foresters can select across the range of diameter classes in the stand and maintain a mosaic of age and diameter classes.

Méthode du Contrôle

Classical European silviculture achieved impressive results with systems such as Henri Biolley's *méthode du contrôle* in Switzerland, in which the number and size of trees harvested were determined by reference to data collected from every tree in every stand measured every 7 years.

While not designed to be applied to boreal mixedwoods, the *méthode du contrôle* is described briefly here to illustrate the degree of sophistication applied by some European foresters to the management of their forests. Development of management techniques that allowed for stand development to be monitored and guided into sustainable paths were in part a response to past experience, particularly in Central European countries, of the negative effects of pure, uniform stands with species often unsuited to the site, which greatly increased the risk of soil degradation and biotic diseases. Increased mortality and decreased increment generated widespread concern, especially after reinforcement by other environmental stresses.

More or less uneven-aged, mixed forests of preponderantly native species, on the other hand, treated along natural lines, have proved to be healthier and more resistant to all kinds of external dangers; and in the long run such stands are more productive and easier to protect.

However, irregular stands of this type are definitely more difficult to manage—new methods and techniques had to be sought particularly for the establishment of inventories, as well as increment control and yield regulation. In Germany, for instance, since the beginning of the nineteenth-century under the influence of G.L. Hartig (1764–1837), yield regulation has been effected almost exclusively by allotment or formula methods based on the conception of the uniform normal forest with a regular succession of cutting areas.

In France, on the other hand, efforts were made to apply another kind of forest management, one that aimed to bring all parts of the forest to a state of highest productive capacity in perpetuity. In 1878, the French forester A. Gurnaud (1825–1898) published a description of a *méthode du contrôle* for determining increment and yield. The method was based on the fact that through careful, selective harvesting, the productivity of the residual stand can be improved, because timber is removed as a cultural operation. In this method, the increment of stands is accurately determined periodically with the object of gradually converting the forest, through selective management and continuous experimentation, to a condition of equilibrium at maximum productive capacity.

Henri Biolley (1858–1939) was the first to apply Gurnaud's inspired ideas to practical forestry. From 1890 on, he managed the forests of his Swiss district according to these principles, devoting himself for almost 50 years to the study of increment and a treatment of stands directed towards the highest production, and proving the practicability of the check method. In 1920, he published this study giving a theoretical basis of management of forests under the check method, describing the procedures to be applied in practice (which he partly developed and simplified), and evaluating the results.

Biolley's pioneering work formed the basis upon which most Swiss forest management practices were later developed, and his ideas have been generally accepted. Today, with the trend of intensifying forest management and productivity in most countries, the ideas and application of careful, continuous treatment of stands with the aid of the volume check method are meeting with ever-growing interest. In Britain and Ireland, for example, there is increased application of Continuous Cover Forestry principles to create permanently irregular structures in many woodlands.

Patch Cut

Row and Broadcast Seeding

Spot and row seeders use less seed that does broadcast ground or aerial seeding but may induce clumping. Row and spot seeding confer greater ability to control seed placement than does broadcast seeding. Also, only a small percentage of the total area needs to be treated.

In the aspen type of the Great Lakes region, direct sowing of the seed of conifers has usually failed. However, Gardner (1980) after trials in Yukon, which included broadcast seeding of white spruce seed at 2.24 kg/ha that secured 66.5% stocking in the Scarified Spring Broadcast treatment 3 years after seeding, concluded that the technique held “considerable promise”.

Seed-tree

An even-aged regeneration method that retains widely spaced residual trees in order to provide uniform seed dispersal across a harvested area. In the seed-tree method, 2-12 seed trees per acre (5-30/ha) are left standing in order to regenerate the forest. They will be retained until regeneration has become established at which point they may be removed. It may not always be economically viable or biologically desirable to re-enter the stand to remove the remaining seed trees. Seed-tree cuts can also be viewed as a clearcut with natural regeneration and can also have all of the problems associated with clearcutting. This method is most suited for light-seeded species and those not prone to windthrow.

Selection Systems

Selection systems are appropriate where uneven stand structure is desired, particularly where the need to retain continuous cover forest for aesthetic or environmental reasons outweighs other management considerations. Selection logging has been suggested as being of greater utility than shelterwood systems in regenerating old-growth Engelmann Spruce Sub-alpine Fir (ESSF) stands in southern British Columbia. In most areas, selection logging favours regeneration of fir more than the more light-demanding spruce. In some areas, selection logging can be expected to favour spruce over less tolerant hardwood species (Zasada 1972) or lodgepole pine.

Shelter Spot Seeding

The use of shelters to improve germination and survival in spot seedings seeks to capture the benefits of greenhouse culture, albeit miniature. The Hakmet seed shelter, for instance, is a semi-transparent plastic cone 8 cm high, with openings of 7 cm diameter in the 7.5 cm diameter base and 17 mm diameter in the 24 mm diameter top. This miniature greenhouse increases air humidity, reduces soil desiccation, and raises air and soil temperatures to levels more favourable to germination and seedling growth than those offered by unprotected conditions. The shelter is designed to break down after a few years of exposure to ultraviolet radiation.

Seed shelters and spring sowing significantly improved stocking compared with bare spot seeding, but sheltering did not significantly improve growth. Stocking of bare seedspots was extremely low, possibly due to smothering of seedlings by abundant broadleaf and herbaceous litter, particularly that from aspen and red raspberry, and exacerbated by strong competition from graminoids and raspberry.

Cone shelters (Cerkon™) usually produced greater survival than unsheltered seeding on scarified seedspots in trials of direct seeding techniques in interior Alaska, and funnel shelters (Cerbel™) usually produced greater survival than unsheltered seeding on non-scarified seedspots. Both shelter types are manufactured by AB Cerbo in Trollhättan, Sweden. Both are made of light-degradable, white, opaque plastic, and are 8 cm high when installed.

White spruce seed was sown in Alaska on a burned site in summer 1984, and protected by white plastic cones on small spots scarified by hand, or by white funnels placed directly into the residual ash and organic material. A group of 6 ravens (*Corvus corax*) was observed in the area about 1 week after sowing was completed in mid-June. Damage averaged 68% with cones and 50% with funnels on an upland area, and 26% with funnels on a floodplain area. Damage by ravens was only 0.13% on unburned but otherwise similar areas.

In seeding trials in Manitoba between 1960 and 1966 aimed at converting aspen stands to spruce–aspen mixedwoods, 1961 scarification in the Duck Mountain Provincial Forest remained receptive to natural seeding for many years.

Shelterwood

In general terms, the shelterwood system is a series of partial cuts that removes the trees of an existing stand over several years and eventually culminates in a final cut that creates a new even-aged stand. It is an even-aged regeneration method that removes trees in a series of three harvests: 1) Preparatory cut; 2) Establishment cut; and 3) Removal cut. The success of practising a shelterwood system is closely related to: 1. the length of the regeneration period, i.e. the time from the shelterwood cutting to the date when a new generation of trees has been established; 2.the quality of the new tree stand with respect to stand density and growth; and 3.the value increment of the shelter trees. Information on the establishment, survival and growth of seedlings influenced by the cover of shelter trees, as well as on the growth of these trees, is needed as a basis for modelling the economic return of practising a shelterwood system. The method's objective is to establish new forest reproduction under the shelter of the retained trees. Unlike the seed-tree method, residual trees alter understory environmental conditions (i.e. sunlight, temperature, and moisture) that influence tree seedling growth. This method can also find a middle ground with the light ambiance by having less light accessible to competitors while still being able to provide enough light for tree regeneration. Hence, shelterwood methods are most often chosen for site types characterized by extreme conditions, in order to create a new tree generation within a reasonable time period. These conditions are valid foremost on level ground sites which are either dry and poor or moist and fertile.

Shelterwood Systems

Shelterwood systems involve 2, 3, or exceptionally more partial cuttings. A final cut is made once adequate natural regeneration has been obtained. The shelterwood system is most commonly applied as a 2-cut uniform shelterwood, first an initial regeneration (seed) cut, the second a final harvest cut. In stands less than 100 years old, a light preparatory cut can be useful. A series of intermediate cuts at intervals of 10–20 years has been recommended for intensively managed stands.

From operational or economic standpoints, however, there are disadvantages to the shelterwood system: harvesting costs are higher; trees left for deferred cutting may be damaged during the regeneration cut or related extraction operations; the increased risk of blowdown threatens the seed source; damage from bark beetles is likely to increase; regeneration may be damaged during the final cut and related extraction operations; the difficulty of any site preparation would be increased; and incidental damage to regeneration might be caused by any site preparation operations.

Single-tree Selection

The single-tree selection method is an uneven-aged regeneration method most suitable when shade tolerant species regeneration is desired. It is typical for older and diseased trees to be removed, thus thinning the stand and allowing for younger, healthy trees to grow. Single-tree selection can be very difficult to implement in dense or sensitive stands and residual stand damage can occur. This method is also disturbs the canopy layer the least out of all other methods.

Spot Seeding

Spot seeding was found to be the most economical and reliable of the direct seeding methods for converting aspen and paper birch to spruce and pine. In the Chippewa National Forest (Lake States), seed-spot sowing of 10 seeds each of white spruce and white pine under 40-year aspen after different degrees of cutting on gave second-season results clearly indicating the need to remove or disturb the forest floor to obtain germination of seeded white spruce and white pine.

Spot seeding of coniferous seed, including white spruce, has had occasional success, but several constraining factors commonly limit germination success: the drying out of the forest floor before the roots of germinants reach underlying moisture reserves; and, particularly under hardwoods, the smothering of small seedlings by snow-pressed leaf litter and lesser vegetation. Kittredge and Gervorkiantz (1929) determined that removal of the aspen forest floor increased germination percentage after the second season in seed spots of both white pine and white spruce, in 4 plots, from 2.5% to 5%, from 8% to 22%, from 1% to 9.5%, and from 0% to 15%.

Spot seeding requires less seed than broadcast seeding and tends to achieve more uniform spacing, albeit sometimes with clumping. The devices used in Ontario for manual spot seeding are the "oil can" seeder, seeding sticks, and shakers. The oil can is a container fitted with a long spout through which a predetermined number of seeds are released with each flick of the seeder.

Strip Cutting

Harvesting cutblocks where only a portion of the trees are to be removed is very different from clearcutting. First, trails must be located to provide access for the felling and skidding/forwarding equipment. These trails must be carefully located to ensure that the trees remaining meet the desired quality criteria and stocking density. Second, the equipment must not damage the residual stand. The further desiderata are outlined by Sauder (1995).

The dearth of seed and a deficiency of receptive seedbeds were recognized as major reasons for the lack of success of clearcut harvesting. One remedy attempted in British Columbia and Alberta has been alternate strip cutting. The greater seed source from uncut trees between the cut strips, and the disturbance to the forest floor within the cut strips could be expected to increase the amount of natural regeneration. Trees were cut to a diameter limit in the cut strips, but large trees in the leave strips often proved too much of a temptation and were cut too, thus removing those trees that would otherwise have been the major source of seed.

An unfortunate consequence of strip thinning was the build-up of spruce beetle populations. Shaded slash from the initial cut, together with an increase in the number of windthrown trees in the leave strips, provided conditions ideally suited to the beetle.

Underplanting

DeLong et al. (1991) suggested underplanting 30- to 40-year-old aspen stands, on the basis of the success of natural spruce in regenerating under stands of such stands: "By planting, spacing can be controlled enabling easier protection of the spruce during stand entry for harvesting of the aspen overstorey".

Variable Retention

A harvesting and regeneration method which is a relatively new silvicultural system that retains forest structural elements (stumps, logs, snags, trees, understory spieces and undisturbed layers of forest floor) for at least one rotation in order to preserve environmental values associated with structurally complex forests.

"Uneven-aged and even-aged methods differ in the scale and intensity of disturbance. Uneven-aged methods maintain a mix of tree sizes or ages within a habitat patch by periodically harvesting individual or small groups of trees, Even-aged methods harvest most or all of the overstory and create a fairly uniform habitat patch dominated by trees of the same age". Even-aged management systems have been the prime methods to use when studying the effects on birds.

Mortality

A survey in 1955–56 to determine survival, development, and the reasons for success or failure of conifer pulpwood plantations (mainly of white spruce) in Ontario and Quebec up to 32 years old found that the bulk of the mortality occurred within the first 4 years of planting, unfavourable site and climate being the main causes of failure.

Advance Growth

Naturally regenerated trees in an understorey prior to harvesting constitute a classic case of good news and bad news. Understorey white spruce is of particular importance in mixedwoods dominated by aspen, as in the B15, B18a, and B19a Sections of Manitoba, and elsewhere. Until the latter part of the last century, white spruce understorey was mostly viewed as money in the bank on a long-term, low interest deposit, with final yield to be realized after slow natural succession, but the resource became increasingly threatened with the intensification of harvesting of aspen. White spruce plantations on mixedwood sites proved expensive, risky, and generally unsuccessful. This prompted efforts to see what might be done about growing aspen and white spruce on the same landbase by protecting existing white spruce advance growth, leaving a range of viable crop trees during the first cut, then harvesting both hardwoods and spruce in the final cut. Information about the understorey component is critical to spruce management planning. The ability of then current harvesting technology and crews employed to provide adequate protection for white spruce understories was questioned by Brace and Bella. Specialized equipment and training, perhaps with financial incentives, may be needed to develop procedures that would confer the degree of protection needed for the system to be feasible. Effective understorey management planning requires more than improved mixedwood inventory.

Avoidance of damage to the understorey will always be a desideratum. Sauder's (1990) paper on mixedwood harvesting describes studies designed to evaluate methods of reducing non-trivial damage to understorey residuals that would compromise their chance of becoming a future crop tree. Sauder concluded that: (1) operational measures that protected residual stems may not unduly increase costs, (2) all felling, conifers and hardwoods, needs to be done in one operation to minimize the entry of the feller-buncher into the residual stand, (3) several operational procedures can reduce understorey damage, some of them without incurring extra costs, and (4) successful harvesting of treatment blocks depends primarily on the intelligent location of skid trails and landings. In summary, the key to protecting the white spruce understorey without sacrificing logging efficiency is a combination of good planning, good supervision, the use of appropriate equipment, and having conscientious, well-trained operators.Even the best plan will not reduce understorey damage unless its implementation is supervised.

New stands need to be established to provide for future supply of commercial white spruce from 150 000 ha of boreal mixedwoods in 4 of Rowe's (1972) regional Forest Sections straddling Alberta, Saskatchewan, and Manitoba, roughly from Peace River AB to Brandon MB. In the 1980s, with harvesting using conventional equipment and procedures, a dramatic increase in the demand for aspen posed a serious problem for the associated spruce understorey. Formerly, white spruce in the understories had developed to commercial size through natural succession under the protection of the hardwoods. Brace articulated a widespread concern: "The need for protection of spruce as a component of boreal mixedwoods goes beyond concern for the future commercial softwood timber supply. Concerns also include fisheries and wildlife habitat, aesthetics and recreation, a general dissatisfaction with cleacutting in mixedwoods and a strong interest in mixedwood perpetuation, as expressed recently in 41 public meetings on forestry development in northern Alberta...".

On the basis of tests of 3 logging systems in Alberta, Brace (1990) affirmed that significant amounts of understorey can be retained using any of those systems provided that sufficient effort is directed towards protection. Potential benefits would include increased short-term softwood timber supply, improved wildlife habitat and cutblock aesthetics, as well as reduced public criticism of previous logging practices. Stewart et al. (2001) developed statistical models to predict the natural establishment and height growth of understorey white spruce in the boreal mixedwood forest in Alberta using data from 148 permanent sample plots and supplementary information about height growth of white spruce regeneration and the amount and type of available substrate. A discriminant model correctly classified 73% of the sites as to presence or absence of a white spruce understorey, based on the amount of spruce basal area, rotten wood, ecological nutrient regime, soil clay fraction, and elevation, although it explained only 30% of the variation in the data. On sites with a white spruce understorey, a regression model related the abundance of regeneration to rotten wood cover, spruce basal area, pine basal area, soil clay fraction, and grass cover ($R^2 = 0.36$). About half the seedlings surveyed grew on rotten wood, and only 3% on mineral soil, and seedlings were 10 times more likely to have established on these substrates than on litter. Exposed mineral soil covered only 0.3% of the observed transect area.

Advance Growth Management

Advance growth management, i.e., the use of suppressed understorey trees, can reduce reforestation costs, shorten rotations, avoid denuding the site of trees, and also reduce adverse impacts on aesthetic, wildlife, and watershed values. To be of value, advance growth must have acceptable species composition and distribution, have potential for growth following release, and not be vulnerable to excessive damage from logging.

The age of advance growth is difficult to estimate from its size, as white that appears to be 2- to 3-year-old may well be more than 20 years old. However, age does not seem to determine the ability of advance growth of spruce to respond to release, and trees older than 100 years have shown rapid rates of growth after release. Nor is there a clear relationship between the size of advance growth and its growth rate when released.

Where advance growth consists of both spruce and fir, the latter is apt to respond to release more quickly than the former, whereas spruce does respond. If the ratio of fir to spruce is large, however, the greater responsiveness to release of fir may subject the spruce to competition severe enough to negate much of the effect of release treatment. Even temporary relief from shrub competition has increased height growth rates of white spruce in northwestern New Brunswick, enabling the spruce to overtop the shrubs.

Site Preparation

Site preparation is any of various treatments applied to a site in order to ready it for seeding or planting. The purpose is to facilitate the regeneration of that site by the chosen method. Site preparation may be designed to achieve, singly or in any combination: improved access, by reducing or rearranging slash, and amelioration of adverse forest floor, soil, vegetation, or other biotic factors. Site preparation is undertaken to ameliorate one or more constraints that would otherwise be likely to thwart the objectives of management. A valuable bibliography on the effects of soil temperature and site preparation on subalpine and boreal tree species has been prepared by McKinnon et al. (2002).

Site preparation is the work that is done before a forest area is regenerated. Some types of site preparation are burning.

Burning

Broadcast burning is commonly used to prepare clearcut sites for planting, e.g., in central British Columbia, and in the temperate region of North America generally.

Prescribed burning is carried out primarily for slash hazard reduction and to improve site conditions for regeneration; all or some of the following benefits may accrue:

a) Reduction of logging slash, plant competition, and humus prior to direct seeding, planting, scarifying or in anticipation of natural seeding in partially cut stands or in connection with seed-tree systems.

b) Reduction or elimination of unwanted forest cover prior to planting or seeding, or prior to preliminary scarification thereto.

c) Reduction of humus on cold, moist sites to favour regeneration.

d) Reduction or elimination of slash, grass, or brush fuels from strategic areas around forested land to reduce the chances of damage by wildfire.

Prescribed burning for preparing sites for direct seeding was tried on a few occasions in Ontario, but none of the burns was hot enough to produce a seedbed that was adequate without supplementary mechanical site preparation.

Changes in soil chemical properties associated with burning include significantly increased pH, which Macadam (1987) in the Sub-boreal Spruce Zone of central British

Columbia found persisting more than a year after the burn. Average fuel consumption was 20 to 24 t/ha and the forest floor depth was reduced by 28% to 36%. The increases correlated well with the amounts of slash (both total and ≥7 cm diameter) consumed. The change in pH depends on the severity of the burn and the amount consumed; the increase can be as much as 2 units, a 100-fold change. Deficiencies of copper and iron in the foliage of white spruce on burned clearcuts in central British Columbia might be attributable to elevated pH levels.

Even a broadcast slash fire in a clearcut does not give a uniform burn over the whole area. Tarrant (1954), for instance, found only 4% of a 140-ha slash burn had burned severely, 47% had burned lightly, and 49% was unburned. Burning after windrowing obviously accentuates the subsequent heterogeneity.

Marked increases in exchangeable calcium also correlated with the amount of slash at least 7 cm in diameter consumed. Phosphorus availability also increased, both in the forest floor and in the 0 cm to 15 cm mineral soil layer, and the increase was still evident, albeit somewhat diminished, 21 months after burning. However, in another study in the same Sub-boreal Spruce Zone found that although it increased immediately after the burn, phosphorus availability had dropped to below pre-burn levels within 9 months.

Nitrogen will be lost from the site by burning, though concentrations in remaining forest floor were found by Macadam (1987) to have increased in 2 of 6 plots, the others showing decreases. Nutrient losses may be outweighed, at least in the short term, by improved soil microclimate through the reduced thickness of forest floor where low soil temperatures are a limiting factor.

The *Picea/Abies* forests of the Alberta foothills are often characterized by deep accumulations of organic matter on the soil surface and cold soil temperatures, both of which make reforestation difficult and result in a general deterioration in site productivity; Endean and Johnstone (1974) describe experiments to test prescribed burning as a means of seedbed preparation and site amelioration on representative clear-felled *Picea/Abies* areas. Results showed that, in general, prescribed burning did not reduce organic layers satisfactorily, nor did it increase soil temperature, on the sites tested. Increases in seedling establishment, survival, and growth on the burned sites were probably the result of slight reductions in the depth of the organic layer, minor increases in soil temperature, and marked improvements in the efficiency of the planting crews. Results also suggested that the process of site deterioration has not been reversed by the burning treatments applied.

Ameliorative Intervention

Slash weight (the oven-dry weight of the entire crown and that portion of the stem < 4 inches in diameter) and size distribution are major factors influencing the forest fire hazard on harvested sites. Forest managers interested in the application of prescribed burning for hazard reduction and silviculture, were shown a method for quantifying the slash load by Kiil (1968). In west-central Alberta, he felled, measured, and weighed 60 white spruce, graphed (a) slash weight per merchantable unit volume against diameter at breast height (dbh), and (b) weight of fine slash (<1.27 cm) also against dbh, and produced a table of slash weight and size distribution on one acre of a hypothetical stand of white spruce. When the diameter distribution of a stand is unknown, an estimate of slash weight and size distribution can be obtained from average stand diameter, number of trees per unit area, and merchantable cubic foot volume. The sample trees in Kiil's study had full symmetrical crowns. Densely growing trees with short and often irregular crowns would probably be overestimated; open-grown trees with long crowns would probably be underestimated.

The need to provide shade for young outplants of Engelmann spruce in the high Rocky Mountains is emphasized by the U.S. Forest Service. Acceptable planting spots are defined as microsites on the north and east sides of down logs, stumps, or slash, and lying in the shadow cast by such material. Where the objectives of management specify more uniform spacing, or higher densities, than obtainable from an existing distribution of shade-providing material, redistribution or importing of such material has been undertaken.

Access

Site preparation on some sites might be done simply to facilitate access by planters, or to improve access and increase the number or distribution of microsites suitable for planting or seeding.

Wang et al. (2000) determined field performance of white and black spruces 8 and 9 years after outplanting on boreal mixedwood sites following site preparation (Donaren disc trenching versus no trenching) in 2 plantation types (open versus sheltered) in southeastern Manitoba. Donaren trenching slightly reduced the mortality of black spruce but significantly increased the mortality of white spruce. Significant difference in height was found between open and sheltered plantations for black spruce but not for white spruce, and root collar diameter in sheltered plantations was significantly larger than in open plantations for black spruce but not for white spruce. Black spruce open plantation had significantly smaller volume (97 cm^3) compared with black spruce sheltered (210 cm^3), as well as white spruce open (175 cm^3) and sheltered (229 cm^3) plantations. White spruce open plantations also had smaller volume than white spruce sheltered plantations. For transplant stock, strip plantations had a significantly higher volume (329 cm^3) than open plantations (204 cm^3). Wang et al. (2000) recommended that sheltered plantation site preparation should be used.

Mechanical

Up to 1970, no "sophisticated" site preparation equipment had become operational in Ontario, but the need for more efficacious and versatile equipment was increasingly recognized. By this time, improvements were being made to equipment originally developed by field staff, and field testing of equipment from other sources was increasing.

According to J. Hall (1970), in Ontario at least, the most widely used site preparation technique was post-harvest mechanical scarification by equipment front-mounted on a bulldozer (blade, rake, V-plow, or teeth), or dragged behind a tractor (Imsett or S.F.I. scarifier, or rolling chopper). Drag type units designed and constructed by Ontario's Department of Lands and Forests used anchor chain or tractor pads separately or in combination, or were finned steel drums or barrels of various sizes and used in sets alone or combined with tractor pad or anchor chain units.

J. Hall's (1970) report on the state of site preparation in Ontario noted that blades and rakes were found to be well suited to post-cut scarification in tolerant hardwood stands for natural regeneration of yellow birch. Plows were most effective for treating dense brush prior to planting, often in conjunction with a planting machine. Scarifying teeth, e.g., Young's teeth, were sometimes used to prepare sites for planting, but their most effective use was found to be preparing sites for seeding, particularly in backlog areas carrying light brush and dense herbaceous growth. Rolling choppers found application in treating heavy brush but could be used only on stone-free soils. Finned drums were commonly used on jack pine–spruce cutovers on fresh brushy sites with a deep duff layer and heavy slash, and they needed to be teamed with a tractor pad unit to secure good distribution of the slash. The S.F.I. scarifier, after strengthening, had been "quite successful" for 2 years, promising trials were under way with the cone scarifier and barrel ring scarifier, and development had begun on a new flail scarifier for use on sites with shallow, rocky soils. Recognition of the need to become more effective and efficient in site preparation led the Ontario Department of Lands and Forests to adopt the policy of seeking and obtaining for field testing new equipment from Scandinavia and elsewhere that seemed to hold promise for Ontario conditions, primarily in the north. Thus, testing was begun of the Brackekultivator from Sweden and the Vako-Visko rotary furrower from Finland.

Mounding

Site preparation treatments that create raised planting spots have commonly improved outplant performance on sites subject to low soil temperature and excess soil moisture. Mounding can certainly have a big influence on soil temperature. Draper et al. (1985), for instance, documented this as well as the effect it had on root growth of outplants (Table 30).

The mounds warmed up quickest, and at soil depths of 0.5 cm and 10 cm averaged 10 and 7 °C higher, respectively, than in the control. On sunny days, daytime surface temperature maxima on the mound and organic mat reached 25 °C to 60 °C, depending on soil wetness and shading. Mounds reached mean soil temperatures of 10 °C at 10 cm depth 5 days after planting, but the control did not reach that temperature until 58 days after planting. During the first growing season, mounds had 3 times as many days with a mean soil temperature greater than 10 °C than did the control microsites.

Draper et al.'s (1985) mounds received 5 times the amount of photosynthetically active radiation (PAR) summed over all sampled microsites throughout the first growing season; the control treatment consistently received about 14% of daily background PAR, while mounds received over 70%. By November, fall frosts had reduced shading, eliminating the differential. Quite apart from its effect on temperature, incident radiation is also important photosynthetically. The average control microsite was exposed to levels of light above the compensation point for only 3 hours, i.e., one-quarter of the daily light period, whereas mounds received light above the compensation point for 11 hours, i.e., 86% of the same daily period. Assuming that incident light in the 100-600 $\mu Em^{-2}s^{-1}$ intensity range is the most important for photosynthesis, the mounds received over 4 times the total daily light energy that reached the control microsites.

Orientation of Linear Site Preparation, e.g., Disk-trenching

With linear site preparation, orientation is sometimes dictated by topography or other considerations, but the orientation can often be chosen. It can make a difference. A disk-trenching experiment in the Sub-boreal Spruce Zone in interior British Columbia investigated the effect on growth of young outplants (lodgepole pine) in 13 microsite planting positions: berm, hinge, and trench in each of north, south, east, and west aspects, as well as in untreated locations between the furrows. Tenth-year stem volumes of trees on south, east, and west-facing microsites were significantly greater than those of trees on north-facing and untreated microsites. However, planting spot selection was seen to be more important overall than trench orientation.

In a Minnesota study, the N–S strips accumulated more snow but snow melted faster than on E–W strips in the first year after felling. Snow-melt was faster on strips near the centre of the strip-felled area than on border strips adjoining the intact stand. The strips, 50 feet (15.24 m) wide, alternating with uncut strips 16 feet (4.88 m) wide, were felled in a *Pinus resinosa* stand, aged 90 to 100 years.

Reforestation

Reforestation is the natural or intentional restocking of existing forests and woodlands that have been depleted, usually through deforestation. Reforestation can be used to rectify or improve the quality of human life by soaking up pollution and dust from the air, rebuild natural habitats and ecosystems, mitigate global warming since forests facilitate biosequestration of atmospheric carbon dioxide, and harvest for resources, particularly timber, but also non-timber forest products.

Tropical tree nursery at Planeta Verde Reforestación S.A.'s plantation in Vichada Department, Colombia

A 15-year-old reforested plot of land

A 21-year-old plantation of red pine in Southern Ontario

The term reforestation is similar to afforestation, the process of restoring and recreating areas of woodlands or forests that may have existed long ago but were deforested or otherwise removed at some point in the past. Sometimes the term *re-afforestation* is used to distinguish between the original forest cover and the later re-growth of forest to an area. Forestation is the establishment of forest growth on areas that either had forest or lacked it. Special tools, e.g. tree planting bars, are used to make planting of trees easier and faster.

Management

A debated issue in managed reforestation is whether or not the succeeding forest will have the same biodiversity as the original forest. If the forest is replaced with only one species of tree and all other vegetation is prevented from growing back, a monoculture forest similar to agricultural crops would be the result. However, most reforestation involves the planting of different seedlots of seedlings taken from the area, often of multiple species. Another important factor is the natural regeneration of a wide variety of plant and animal species that can occur on a clear cut. In some areas the suppression of forest fires for hundreds of years has resulted in large single aged and single species forest stands. The logging of small clear cuts and/or prescribed burning, actually increases the biodiversity in these areas by creating a greater variety of tree stand ages and species.

For Harvesting

Reforestation need not be only used for recovery of accidentally destroyed forests. In some countries, such as Finland, many of the forests are *managed* by the wood products and pulp and paper industry. In such an arrangement, like other crops, trees are planted to replace those that have been cut. In such circumstances, the industry can cut the trees in a way to allow easier reforestation. The wood products industry systematically replaces many of the trees it cuts, employing large numbers of summer workers for tree planting work. For example, in 2010, Weyerhaeuser reported planting 50 million seedlings. However replanting an old-growth forest with a plantation is not replacing the old with the same characteristics in the new.

In just 20 years, a teak plantation in Costa Rica can produce up to about 400 m^3 of wood per hectare. As the natural teak forests of Asia become more scarce or difficult to obtain, the prices commanded by plantation-grown teak grows higher every year. Other species such as mahogany grow more slowly than teak in Tropical America but are also extremely valuable. Faster growers include pine, eucalyptus, and *Gmelina.*

Reforestation, if several indigenous species are used, can provide other benefits in addition to financial returns, including restoration of the soil, rejuvenation of local flora and fauna, and the capturing and sequestering of 38 tons of carbon dioxide per hectare per year.

The reestablishment of forests is not just simple tree planting. Forests are made up of a community of species and they build dead organic matter into soils over time. A major tree-planting program could enhance the local climate and reduce the demands of burning large amounts of fossil fuels for cooling in the summer.

For Climate Change Mitigation

Forests are an important part of the global carbon cycle because trees and plants absorb carbon dioxide through photosynthesis. By removing this greenhouse gas from the air, forests function as terrestrial carbon sinks, meaning they store large amounts of carbon. At any time, forests account for as much as double the amount of carbon in the atmosphere. Even as more anthropogenic carbon is produced, forests remove around three billion tons of anthropogenic carbon every year. This amounts to about 30% of all carbon dioxide emissions from fossil fuels. Therefore, an increase in the overall forest cover around the world would tend to mitigate global warming.

There are four major strategies available to mitigate carbon emissions through forestry activities: increase the amount of forested land through a reforestation process; increase the carbon density of existing forests at a stand and landscape scale; expand the use of forest products that will sustainably replace fossil-fuel emissions; and reduce carbon emissions that are caused from deforestation and degradation.

Achieving the first strategy would require enormous and wide-ranging efforts. However, there are many organizations around the world that encourage tree-planting as a way to offset carbon emissions for the express purpose of fighting climate change. For example, in China, the Jane Goodall Institute, through their Shanghai Roots & Shoots division, launched the Million Tree Project in Kulun Qi, Inner Mongolia to plant one million trees to stop desertification and help curb climate change. China has used 24 billion metres squared of new forest plantation and natural forest regrowth to offset 21% of Chinese fossil fuel emissions in 2000. In Java, Indonesia each newlywed couple is to give whoever is sermonizing their wedding 5 seedlings to combat global warming. Each couple that wishes to have a divorce has to give 25 seedlings to whoever divorces them.

The second strategy has to do with selecting species for tree-planting. In theory, planting any kind of tree to produce more forest cover would absorb more carbon dioxide from the atmosphere. On the other hand, a genetically modified tree specimen might grow much faster than any other regular tree. Some of these trees are already being developed in the lumber and biofuel industries. These fast-growing trees would not only be planted for those industries but they can also be planted to help absorb carbon dioxide faster than slow-growing trees.

Extensive forest resources placed anywhere in the world will not always have the same impact. For example, large reforestation programs in boreal or subarctic regions have a limited impact on climate mitigation. This is because it substitutes a bright snow-dominated region that reflects the sunlight with dark forest canopies. A study from the National Center for Atmospheric Research in Boulder, Colorado, USA, found that trees in temperate latitudes have a net warming effect on the atmosphere. The heat that dark leaves release without absorbing outweighs the carbon they sequester. On the other hand, a positive example would be reforestation projects in tropical regions, which would lead to a positive biophysical change such as the formation of clouds. These clouds would then reflect the sunlight, creating a positive impact on climate mitigation.

There is an advantage to planting trees in tropical climates with wet seasons. In such a setting, trees have a quicker growth rate because they can grow year-round. Trees in tropical climates have, on average, larger, brighter, and more abundant leaves than non-tropical climates. A study of the girth of 70,000 trees across Africa has shown that tropical forests are soaking up more carbon dioxide pollution than previously realized. The research suggests almost one fifth of fossil fuel emissions are absorbed by forests across Africa, Amazonia and Asia. Simon Lewis, a climate expert at the University of Leeds, who led the study, said: "Tropical forest trees are absorbing about 18% of the carbon dioxide added to the atmosphere each year from burning fossil fuels, substantially buffering the rate of change."

It is also important to deal with the rate of deforestation. At this point, there are 13 billion metres squared of tropical regions that are deforested every year. There is potential for these regions to reduce rates of deforestation by 50% by 2050, which would be a huge contribution to stabilize the global climate.

Incentives

Some incentives for reforestation can be as simple as a financial compensation. Streck and Scholz (2006) explain how a group of scientists from various institutions have developed a compensated reduction of deforestation approach which would reward developing countries that disrupt any further act of deforestation. Countries that participate and take the option to reduce their emissions from deforestation during a committed period of time would receive financial compensation for the carbon dioxide emissions that they avoided. To raise the payments, the host country would issue government bonds or negotiate some kind of loan with a financial institution that would want to take part in the compensation promised to the other country. The funds received by the country could be invested to help find alternatives to the extensive cutdown of forests. This whole process of cutting emissions would be voluntary, but once the country has agreed to lower their emissions they would be obligated to reduce their emissions. However, if a country was not able to meet their obligation, their target would get added to their next commitment period. The authors of these proposals see this as a solely government-to-government agreement; private entities would not participate in the compensation trades.

Examples

Forest regrowth in Mount Baker-Snoqualmie National Forest, Washington state, USA

Sub-saharan Africa

Plans to plant a nine-mile width of trees on the Southern Border of the Sahara Desert.

The Great Green Wall initiative is a pan-African proposal to "green" the continent from west to east in order to battle desertification. It aims at tackling poverty and the degradation of soils in the Sahel-Saharan region, focusing on a strip of land of 15 km (9 mi) wide and 7,500 km (4,750 mi) long from Dakar to Djibouti.

Canada

In Canada, overall forest cover is increasing over the last decades.

China

In China, extensive replanting programs have existed since the 1970s. Programs have had overall success. The forest cover has increased from 12% of China's land area to 16%. However, specific programs have had limited success. The "Green Wall of China", an attempt to limit the expansion of the Gobi Desert is planned to be 2,800 miles (4,500 km) long and to be completed in 2050. China plans to plant 26 billion trees in the next decade that is two trees for every Chinese citizen per year. China requires that students older than 11 years old plant one tree a year until their high school graduation.

Germany

Reforestation is required as part of the federal forest law. 31% of Germany is forested, according to the second forest inventory of 2001–2003. The size of the forest area in Germany increased between the first and the second forest inventory due to forestation of degenerated bogs and agricultural areas.

India

Jadav Payeng had received national awards for reforestation efforts, known as the "Molai forest".

Japan

Mass environmental and human-body pollution along with relating deforestation, water pollution, smoke damage, and loss of soils caused by mining operations in Ashio, Tochigi became the first environmental social issue in Japan, efforts by Shōzō Tanaka had grown to large campaigns against copper operation. This led to the creation of 'Watarase Yusuichi Pond', to settle the pollution which is a Ramsar site today. Reforestation was conducted as a part of afforestation due to inabilities of self-recovering by the natural land itself due to serious soil pollution and loss of woods consequence in loss of soils for plants to grow, thus needing artificial efforts involving introducing of healthy soils from outside. Starting since in 1897, about 50% of once bald mountains backed to green.

Lebanon

For thousands of years Lebanon was covered by forests, one particular species of interest, *Cedrus libani* or Lebanon Cedar was exceptionally valuable timber species. Virtually every ancient culture that shared the Mediterranean Sea harvested these trees, the Ancient Egyptians, the Greeks, the Roman Empire, early Christians, the Muslims, Persians, Assyrians and Babylonians. Extensive deforestation has occurred, with only small remnants of the original forests surviving. Deforestation has been particularly severe in Lebanon and on Cyprus. The Lebanon Reforestation Initiative aims to restore Lebanon's native forests. Projects financed locally and by international charity are performing extensive reforestation of cedar being carried out in the Mediterranean region, particularly in Lebanon and Turkey, where over 50 million young cedars are being planted annually.

Philippines

The Philippines increased its forest area to 240,000 hectares (593,053 acre) per year from 2010 to 2015. In 2011-2015 period, reforestation through the government's National Greening Program covered 1,258,692 hectares of land, which is 105 percent more than the 1,200,000 hectares planting target for the same period. An estimated eight million hectares of land nationwide need reforestation and rehabilitation, according to the Forest Management Bureau.

United States

It is the stated goal of the US Forest Service to manage forest resources sustainably. This includes reforestation after timber harvest, among other programs.

Organizations

Trees for the Future has assisted more than 170,000 families, in 6,800 villages of Asia, Africa and the Americas, to plant over 35 million trees.

Wangari Maathai, 2004 Nobel Peace Prize recipient, founded the Green Belt Movement which planted over 47 million trees to restore the Kenyan environment.

Shanghai Roots & Shoots, a division of the Jane Goodall Institute, launched The Million Tree Project in Kulun Qi, Inner Mongolia to plant one million trees to stop desertification and alleviate global warming.

Criticisms

Reforestation competes with other land uses such as food production, livestock grazing and living space for further economic growth. Reforestation often has the tendency to create large fuel loads, resulting in significantly hotter combustion than fires involving low brush or grasses. Reforestation can divert large amounts of water from other activities. Reforesting sometimes results in extensive canopy creation that prevents growth of diverse vegetation in the shadowed areas and generating soil conditions that hamper other types of vegetation. Trees used in some reforesting efforts (e.g., *Eucalyptus globulus*) tend to extract large amounts of moisture from the soil, preventing the growth of other plants.

There is also the risk that through a forest fire or insect outbreak much of the stored carbon in a reforested area could make its way back to the atmosphere. Reduced harvesting rates and fire suppression have caused an increase in the forest biomass in the western United States over the past century. This causes an increase of about a factor of four in the frequency of fires due to longer and hotter dry seasons.

Selection Cutting

The results of selective cutting of Ponderosa Pine

Selection cutting is the silvicultural practice of harvesting trees in a way that moves a forest stand towards an uneven-aged or all-aged condition, or 'structure'. Using stocking models derived from the study of old growth forests, '"Selection cutting"', also known as 'selection system', or 'selection silviculture', manages the establishment, continued growth and final harvest of multiple age classes (usually three, but 5 or even 10 are possible) of trees within a stand. A closely related approach to forest management is Continuous Cover Forestry (CCF), which makes use of selection systems to achieve a permanently irregular stand structure.

Selection cutting or systems are generally considered to be more challenging to implement and maintain than even-aged silviculture, due to the difficulty of managing multiple age classes in a shared space, but there are significant ecological benefits associated with it. Uneven-aged stands generally exhibit higher levels of vertical structure (key for many species of birds and mammals), have higher levels of carbon sequestration, and produce a more constant flow of market and non-market forest resources than even-aged stands.. Although a forest composed of many stands with varied maturity ages maybe comparable, this would be at the forest rather than the stand level. This silvicultural method also protects forest soils from the adverse effects of many types of even-aged silviculture, including nutrient loss, erosion and soil compaction and the rapid loss of organic material from a forested system. Selection silviculture is especially adept at regenerating shade-tolerant species of trees (those able to function under conditions of low solar energy, both cooler and less light), but can also be modified to suit the regeneration and growth of intolerant and mid-tolerant species. This is one of many different ways of harvesting trees. Selection cutting as a silvicultural system can be modified in many ways and would be so done be a forester to take into account varied ownership goals, local site conditions and the species mix found from past forest conditions.

Confusing Term

Selection cutting is often (sometimes deliberately) confused with "selective" cutting, a term synonymous with the practice of highgrading (the removal of the most economically profitable trees in a forest, often with a disregard for the future of the residual stand). Often the latter term is used by foresters or loggers to imply the former (which has a generally positive connotation in forestry circles) and mislead landowners into stripping their woodlot of its most valuable timber. Used correctly, the term 'selection cutting', 'selection system', or 'selection silviculture' implies the implementation of specific silvicultural techniques—usually either 'single tree selection', 'group selection' or a combination of the two—to create an uneven-aged or all-aged condition in a forest stand, one more akin to a late successional or 'climax' condition.

Partly as a result of such confusion, the term *Plenterwald*, which is the German term for selection cutting, is being more commonly used as the standard term in English. Increasingly, especially in Britain, Ireland and elsewhere in Europe, the term Continuous Cover Forestry (CCF) has been adopted to embrace an approach to stand management that most often employs selection systems to achieve a permanently irregular stand structure.

Single-tree Selection

The most common type of selection system is single-tree selection, in which scattered individual trees of multiple age classes, whose canopies are not touching, are harvested. This type of selection system generally produces small canopy openings especially conducive to the establishment and growth of shade tolerant tree species.

Group Selection

Another variation of selection silviculture is group selection. Under this system, a number of 'groups', or small openings created by the removal of several adjacent trees, are created in complement to the harvest of scattered individual trees. If the groups created are large enough, and if seed-bed conditions are favorable, this can allow species which are intolerant of shade to regenerate. Group selection is designed to mimic larger, multi-tree mortality events, which in some environments may represent natural disturbance regimes.

The maximum size of a group (before it becomes a patch, or clearcut) is debatable. Some say it may be up to 2 acres (0.8 hectares) in size, whereas others limit it to a maximum of 0.5 acres (0.1 hectares).

In any case Plenterwald can operate in a small areas of 1/3 - 1/2 hectare, whereas other systems need a bigger area. Behind this is the philosophical idea that a stand should be balanced (that is equal amounts of land cover for each age class) in the same way that a forest would be balanced under a clear cut régime (that is stands collectively are balanced on yield flow). The reasoning is based on the Normalwald concept, which is a model of a forest over 100 years that will produce an amount of money that is consistent over time with treatments being consistent over time rather than big expenses or big profits at one time and low expenses and low profits at another.

Care need to be taken to avoid epicormic shoots growing on trunks of surrounding trees such that they lead to knotty wood, if timber production is desired. It is also challenging to visualize the groups with cuttings over time.

Implementing a Selection System

In North America, trees are selected for harvest in a selection system with reference to the Arbogast Method (named after the method's creator). This is also known as the *BDq* method. Under this method, a harvest is specified by defining a residual basal area (*B*), a maximum diameter (*D*), and a q-ratio (*q*). The q-ratio is the ratio of the number of trees in a diameter class to the number of trees in the next larger class. Typically diameter classes are either 4 centimeters or 2 inches.

When the Q is plotted on semi-log paper it gives a straight slope for uneven aged stands. However, in reality this slope can be seen to vary from what is called an S-curve in old growth forests to cut off the older trees giving a reverse-J curve in a managed stand. (Forester owners usually do not want lots of old moldy trees as they want to sell the timber for use in building etc.) The curve is also an ideal curve and there may be variations to some extent, particularly in earlier number of trees where there are many more seedlings and saplings than the model Q-ratio would suggest.

Given the BDq, a curve representing the state of the residual stand is computed. This curve is compared to the inventory data from a stand, specifically the curve of the diameter classes of the trees in the stand (diameter is used as a surrogate for age and thus called an age class even though strictly it should be a size class) against the number of trees in each diameter (age) class. The comparison of these two curves tells the forester how many trees of each age-class should remain in the stand. The excess trees are marked for harvest. If there are too few trees in a class the forester will use their judgement to not take so many trees from neighboring classes.

The goal of the use of a BDq curve is to ensure the continued development of trees in each age class, and the continued availability of mature timber to harvest on a relatively short cutting cycle (8–15 years). Longer cutting cycles may be used depending on species mix, silvicultural goals and if the aim is amenity or economic forestry in respect to the land.

Following this method with well performed forest inventories should see the right amount of cutting. However, reality has shown about a third of forests are overcut and a third are undercut. It appears that the model also departs from reality in many cases, and so cannot be solely relied on. The judgement of an experienced forester is also needed.

Controlled Burn

An experimental burn in Canada; note the measuring equipment in front.

Firing the woods in a South Carolina forest with a custom made driptorch mounted on an ATV. The device spits flaming fuel oil from the side, instantly igniting the leaf litter.

A prescribed burn in a *Pinus nigra* stand in Portugal

Controlled burning of a field outside of Statesboro, Georgia, United States in preparation for spring planting

An aerial view of a controlled burn in Helderberg Nature Reserve in South Africa bordering the city of Cape Town. In South Africa controlled burns are important for maintaining the ecological health of indigenous fynbos as well as reducing the intensity of future burns.

Japanese controlled burn (Hokkaido)

Controlled or prescribed burning, also known as hazard reduction burning (HRB), backfire, or swailing, is a technique sometimes used in forest management, farming, prairie restoration or greenhouse gas abatement. Fire is a natural part of both forest and grassland ecology and controlled fire can be a tool for foresters. Hazard reduction or controlled burning is conducted during the cooler months to reduce fuel buildup and decrease the likelihood of serious hotter fires. Controlled burning stimulates the germination of some desirable forest trees, and reveals soil mineral layers which increases seedling vitality, thus renewing the forest. Some cones, such as those of Lodgepole Pine and Sequoia, are serotinous, as well as many chaparral shrubs, meaning they require heat from fire to open cones to disperse seeds.

In industrialized countries, controlled burning is usually overseen by fire control authorities for regulations and permits. The party responsible must delineate the intended time and place. Obtaining a permit may not limit liability if the fire burns out of control.

History

There are two basic causes of wildfires. One is natural (lightning) and the other is people. Controlled burns have a long history in wildland management. Pre-agricultural societies used fire to regulate both plant and animal life. Fire history studies have documented periodic wildland fires ignited by indigenous peoples in North America and Australia.

Fires, both naturally caused and prescribed, were once part of natural landscapes in many areas. Studies have shown that between the mid Holocene and the 17th century AD, wildland fires annually burned between 45.0% and 87.5% of present-day California's total land, for example. These practices ended in the early 20th century when US fire policies were enacted with the goals of suppressing all fires. Since 1995, the US Forest Service has slowly incorporated burning practices into its forest management policies.

Back Burning

Back burning is a technique utilized in controlled burning and during wildfire events. While controlled burns utilize back burning during planned fire events to create a "black line", back burning or backfiring is also done to stop a wildfire that is already in progress. Back burning is a way of reducing the amount of flammable material during a controlled burn or wildfire by starting small fires along a man made or natural firebreak in front of a main fire front. It is called back burning because the small fires are designed to 'burn back towards the main fire front' and are usually burning and traveling against ground level winds. Back burning reduces the amount of fuel that's available to the main fire by the time that it reaches the burnt area. Firebreaks are often used as an anchor point to start a line of fires along natural or manmade features such as a river, road or a bulldozed clearing.

Forest Use

Another consideration is the issue of fire prevention. In Florida, during the drought in 1995, catastrophic wildfires burned numerous homes. But forestry managers in the Florida Division of Forestry noted that the underlying problem was previous cessation of controlled burning, resulting from complaints by homeowners. Each year additional leaf litter and dropped branches increased the likelihood of a hot and uncontrollable fire.

Controlled burns are sometimes ignited using a tool known as the driptorch, which allows a steady stream of flaming fuel to be directed to the ground as needed. Variations on the driptorch can be used such as the *helitorch*, which is mounted on a helicopter, or other improvised devices such as mounting a driptorch-like device on the side of a vehicle. A pyrotechnic device known as a fusee can be used for ignition in nearby fuels while a Very pistol can be for fuels farther away.

For the burning of slash, waste materials left over from logging, there are several types of controlled burns. Broadcast burning is the burning of scattered slash over a wide area. Pile burning is gathering up the slash into piles before burning. These burning piles may be referred to as bonfires. High temperatures can harm the soil, damaging it physically, chemically or sterilizing it. Broadcast burns tend to have lower temperatures and will not harm the soil as much as pile burning, though steps can be taken to treat the soil after a burn. In lop and scatter burning, slash is left to compact over time, or is compacted with machinery. This produces a lower intensity fire, as long as the slash is not packed too tightly. However, soil may be damaged if machinery is used to compress the slash.

Controlled burning reduces fuels, may improve wildlife habitat, controls competing vegetation, improves short term forage for grazing, improves accessibility, helps control tree disease, and perpetuates fire dependent species. In mature longleaf pine forest, it helps maintain habitat for endangered red-cockaded woodpeckers in their sandhill and flatwoods habitats. Fire is also felt to be a crucial element of the recovery of the threatened Louisiana pine snake in the longleaf pine forests of central Louisiana and eastern Texas. However, many scientists disagree with such a simplistic approach, and indicate that each forest must be assessed on its own merit.

In the wild, many trees depend on fire as a successful way to clear out competition and release their seeds. In particular, the Giant sequoia depends on fire to reproduce: the cones of the tree open after a fire releases their seeds, the fire having cleared all competing vegetation. Due to fire suppression efforts during the early and mid 20th century, low-intensity fires no longer occurred naturally in many groves, and still do not occur in some groves today. The suppression of fires also led to ground fuel build-up and the dense growth which posed the risk of catastrophic wildfires. It wasn't until the 1970s that the National Park Service began systematic fires for the purpose of new seed growth. Eucalyptus regnans or Mountain Ash of Australia also depends on fire but in a different fashion. They carry their seeds in capsules which can deposit at any time of the year. Being flammable, during a fire the capsules drop nearly all of them and the fire consumes the Eucalypt adults, but most of the seeds survive using the ash as a source of nutrients; at their rate of growth, they quickly dominate the land and a new Eucalyptus forest grows.

Agricultural Use

In addition to forest management, controlled burning is also used in agriculture. In the developing world, this is often referred to as slash and burn. In industrialized nations, it is seen as one component of shifting cultivation, as a part of field preparation for planting. Often called field burning, this technique is used to clear the land of any existing crop residue as well as kill weeds and weed seeds. Field burning is less expensive than most other methods such as herbicides or tillage, but because it does produce smoke and other fire-related pollutants, its use is not popular in agricultural areas bounded by residential housing.

In the United States, field burning is a legislative and regulatory issue at both the federal and state levels of government.

Controversy

In Oregon, field burning has been widely used by grass seed farmers as a method for clearing fields for the next round of planting, as well as revitalizing serotinous grasses that require fire in order to grow seed again. The Oregon Department of Environmental Quality began requiring a permit for farmers to burn their fields in 1981, but the requirements became stricter in 1988 following a multi-car collision in which smoke from field burning near Albany, Oregon, obscured the vision of drivers on Interstate 5, leading to a 23-car collision in which 5 people died and 37 were injured. This resulted in more scrutiny of field burning and proposals to ban field burning in the state altogether.

In the European Union burning crop stubble after harvest is used by farmers for plant health reasons under several restrictions in the cross compliance regulations.

Heathland Use

The controlled burning or swailing of heathland is used in the United Kingdom and other countries as a conservation tool. In Scotland it is known as "muirburn". Often it is used as a tool for creating fire breaks to reduce the risk of dangerous outbreaks but is also an important mechanism for preventing succession to woodier vegetation and plays an important role in the life cycle of heather species. Red grouse eat heather shoots and as grouse shooting has long been an expensive sport the landowners need to ensure a good crop of grouse so each year parts of the moor are burned of old heather to cause fresh shoots.

Greenhouse Gas Abatement

Controlled burns on Australian savannas can result in an overall reduction of greenhouse gas emissions. One working example is the West Arnhem Fire Management Agreement, started to bring "strategic fire management across 28,000 square kilometres (11,000 sq mi) of Western Arnhem Land" to partially offset greenhouse gas emissions from a liquefied natural gas plant in Darwin, Australia. Deliberately starting controlled burns early in the dry season results in a mosaic of burnt and unburnt country which reduces the area of stronger, late dry season fires. Also known as "patch burning". To minimise the impact of smoke, burning should be restricted to daylight hours whenever possible.

Northern California fire crews start a backfire to stop the Poomacha fire from advancing westward.

Forest Protection

Forest security in Lithuania

Forest protection is the preservation or improvement of a forest threatened or affected by abuse, or out of balance nature.

Thus forest protection also has a legal status and rather than protection from only people damaging the forests is seen to be broader and include forest pathology too. Thus due to this the different emphases around the world paradoxically suggest different things for forest protection.

In German speaking countries forest protection would focus on the biotic and abiotic factors that are non-crime related. A protected forest is not the same as a protection forest. These terms can lead to some confusion in English, although they are clearer in other languages. As a result, reading English literature can be problematic for non-experts due to localization and conflation of meanings.

The types of man induced abuse that forest protection seeks to prevent include:

- Aggressive or unsustainable farming and logging
- Expanding city development caused by population explosion and the resulting urban sprawl

There is considerable debate over the effectiveness of forest protection methods. Enforcement of laws regarding purchased forest land is weak or non-existent in most parts of the world. In the increasingly dangerous South America, home of major rainforests, officials of the Brazilian National Agency for the Environment (IBAMA) have recently been shot during their routine duties.

Land Purchase

One simple type of forest protection is the purchasing of land in order to secure it, or in order to plant trees (afforestation). It can also mean forest management or the designation of areas such as natural reservoirs which are intended to be left to themselves. However, merely purchasing a piece of land does not prevent it from being used by others for poaching and illegal logging.

On Site Monitoring

A better way to protect a forest, particularly old growth forests in remote areas, is to obtain a part of it and to live on and monitor the purchased land. Even in the United States, these measures sometimes don't suffice because arson can burn a forest to the ground, leaving burnt areas free for different use.

Another issue about living on purchased forest-land is that there may not be a suitable site for a standard home without clearing land, which defies the purpose of protection. Alternatives include building a treehouse or an earthhouse. This is being done currently by indigenous people in South America to protect large reservoirs. In former times, North American Native Americans used to live in tipies or mandan earthhouses, which also require less land. An undertaking to develop modern treehouses is being taken by a company from Germany called "TrueSchool treehouses".

Other Methods of Protection

A number of less successful methods of forest protection have been tried, such as the trade in certified wood. Protecting a small section of land in a larger forest may also have limited value. For example, tropical rainforests can die if they decrease in size, since they are dependent on the moist microclimate which they create. There is an excellent article in National Geographic October issue concerning redwood forest in California and their effort to maintain forest and rainforest.

A compromise is to conduct agriculture and stock farming, or sustainable wood management. This ascribes different values to forest land and farmland, for which many areas are clear felled.

'Neighborhood Leakage'

Two conflicting studies on the idea that protecting forests only relocates deforestation. This is called neighborhood leakage. According to the paradox of forest protection protected areas such as rural settlements near protected zones grew at twice the rate of those elsewhere. The IUCN implements such protocols that protect over 670 eco-regions. 46% of the eco-regions had less than 10% forest protection. Which means that these areas are not being monitored as they should and the protection is not working. Considering forest protection within global priority areas was unsatisfactory. An example given was that the average protection of 8.4% in biodiversity hotspots. Results have policy relevance in terms of the target of the Convention on Biological Diversity, reconfirmed in 2008, to conserve in an effective manner that "at least 10% of each of the world's forest types".

Urban Forests

A recent discovery in Europe relating to forest protection is that urban areas have forests of their own. Many cities have tens of thousands of trees which constitute forests. In addition the air in the cities is lately becoming better, providing conditions favorable for small associated species such as mosses and lichens.

Certified Wood

A managed Forest on San Juan Island in Washington (U.S. state).

Certified wood and paper products come from responsibly managed forests – as defined by a particular standard. With third-party forest certification, an independent organization develops standards of good forest management, and independent auditors issue certificates to forest operations that comply with those standards.

Requirements

Forest certification programs typically require that forest management practices conform to existing laws. Other basic requirements or characteristics of forest certification programs include:

Basic requirements of credible forest certification programs include:

- Protection of biodiversity, species at risk and wildlife habitat; sustainable harvest levels; protection of water quality; and prompt regeneration (e.g., replanting and reforestation).
- Third-party certification audits performed by accredited certification bodies.
- Publicly available certification audit summaries.
- Multi-stakeholder involvement in a standards development process.
- Complaints and appeals process.

Programs

Today there are more than 50 certification programs worldwide addressing the many types of forests and tenures around the world. The two largest international forest certification programs are the Forest Stewardship Council (FSC) and the Programme for the Endorsement of Forest Certification (PEFC).

The PEFC is the largest certification framework in terms of forest area, with approximately two-thirds of the total certified area. The FSC program is the fastest growing.

Third-party forest certification was pioneered in the early 1990s by the FSC, a collaboration between environmental NGOs, forest product companies and social interests. Competing systems quickly emerged throughout the world. Some commentators, including Jared Diamond, have suggested that many competing standards were set up by logging companies specifically aiming to confuse consumers with less rigorously enforced but similarly named competing standards.

United States and Canada

In the United States and Canada, there are a number of forest certification programs. Three of these programs are endorsed by the PEFC. They are the American Tree Farm System (ATFS), the Canadian Standards Association's Sustainable Forest Management Standard and the Sustainable Forestry Initiative (SFI) Program. ATFS is applicable only in the United States; the Canadian Standards Association SFM Standard is applicable only in Canada. SFI is applicable to both the United States and Canada. The FSC, program is applied throughout North America. SFI is the world's largest regional forest certification program in terms of total certified forest area.

The National Association of State Foresters in the USA passed a resolution in 2008 that supports *all* of the forest certification systems used in the USA and recognized the value of their differences: "... the ATFS, FSC, and SFI systems include the fundamental elements of credibility and make positive contributions to forest sustainability.... No certification program can credibly claim to be 'best', and no certification program that promotes itself as the only certification option can maintain credibility. Forest ecosystems are complex and a simplistic 'one size fits all' approach to certification cannot address all sustainability needs.".

The Canadian Council of Forest Ministers issued a statement in 2008 on forest certification standards in Canada, which said: "In Canada, each jurisdiction's forest laws, policies and administrative requirements comprise an over-arching framework that fully characterizes what sustainable forest management (SFM) means in that jurisdiction, and what actions may take place on public and/or private forest land. Governments in Canada support third-party forest certification as a tool to demonstrate the rigor of Canada's forest management laws, and to document the country's world-class sustainable forest management record. The forest management standards of the Canadian Standards Association (CSA), the FSC and the Sustainable Forestry Initiative (SFI) are all used in Canada. Governments in Canada accept that these standards demonstrate, and promote the sustainability of forest management practices in Canada."

Chain-of-custody Certification

Chain of Custody certification tracks the certified material through the production process – from the forest to the consumer, including all successive stages of processing, transformation, manufacturing and distribution. It provides evidence that certified material in a certified product originates from certified forests.

The United Nations reports that between January 2009 and May 2010, the total number of PEFC and FSC chain-of-custody certificates issued worldwide increased by 88% for a total of 23,717 certificates (this does not include SFI certificates).

Future expansion

Forest certification is a voluntary process about 10% of the world's forest under at least one certification program. Customers that choose to buy certified products are supporting land managers, land owners and forest product companies that have made a commitment to meeting the standards of forest certification.

Third-party forest certification is a useful tool for those seeking to purchase paper and wood products that come from forests that are well-managed and use materials that are legally harvested. Incorporating third-party certification into forest product buying practices can be a centerpiece for responsible wood and paper purchasing policies that include factors such as the protection of sensitive forest values, thoughtful material selection and efficient use of products.

The 2009-2010 United Nations Market Review reported that companies that produced or traded in certified forest products often had a market advantage during the 2008-2009 recession because, in a buyers' market, buyers could be more selective in choosing their sources of supply. The report cites four demand drivers for certification:

- Paper, publishing, printing and packaging – commitments to increase the use of responsible paper sources by large publishers such as Time Inc. has probably been the most significant factor driving growth in forest and chain-of-custody certification.
- Green public procurement – governments such as the UK and the Netherlands have adopted green timber procurement policies, including recognition of FSC and PEFC endorsed programs. An example is the UK's Central Point of Expertise on Timber (set up by the Department of Environment, Food and Rural Affairs and operated by ProForest).
- Green building – standards for green building incentivize and reward the use of certified wood products.
- Illegal logging – new legislation designed to minimize the risk of illegal wood entering supply chains such as the amended Lacey Act in the United States has created a strong incentive to demand independently certified wood that can address illegal logging concerns.

The World Resources Institute, in partnership with the Environmental Investigation Agency, released a fact sheet designed to answer some of the frequently asked questions about the Lacey Act, which was amended in 2008 to ban commerce in illegally sourced plants and their products—including timber, wood, and paper products. The fact sheet says forest certification is a very good approach for demonstrating due care by showing government and customers that a company has taken proactive steps to eliminate illegal wood or plant material from its supply chain. Certification does not relieve importers of the requirement to submit appropriate import declaration information to U.S. government agencies.

Forest Integrated Pest Management

Forest integrated pest management or Forest IPM is the practice of monitoring and managing pest and environmental information with pest control methods to prevent pest damage to forests and forest habitats by the most economical means. Forest IPM practices vary from region to region and particularly by state, according to the habitat and forests present.

In the state of Vermont, two common pests are of particular significance, the Hemlock Woolly Adelgid or HWA, and thrips.

Tropical Rainforest Conservation

Conservation

Right now, people are conserving the Tropical Rain Forests by ecotourism and rehabilitation. Ecotourism is giving people tours of the forest and showing them what we are losing by cutting them down. We are helping them even more by rebuilding and restarting forests in certain areas.

By speaking with the local people living in, and around, the rainforest, conservationists can learn information that would allow them to best focus their conservation efforts.

Another way conservation has become the most economically beneficial option is through carbon credits. Under the Kyoto Protocol, countries must reduce their emissions of Carbon Dioxide by 5% below the 1990 levels before 2012. Countries can meet their mandatory cuts in emissions by offsetting some of those emissions some other way. Through conservation or reforestation of the rainforest, countries can receive credits.

Some worldwide companies have stated publicly that they won't buy products that come from recently cleared areas of the rainforest (beef from the rainforests often come from areas where forests have been destroyed.

It's important to conserve the rainforest because many resources for things we use everyday come from the rainforest, including rubber for tires and spices such as cinnamon and many other common items. The rainforest also needs to be conserved because the earth needs trees to take in carbon dioxide. About 1/4 of the world's greenhouse gas emissions come from deforestation

References

- Nathan, Susan (2005). The Other Side of Israel: My Journey Across the Jewish/Arab Divide. New York: Nan A. Talese. pp. 130–131. ISBN 978-0-385-51456-9.
- Kohm, K. A, and Franklin, J. F., Creating a forestry for the 21st century: the science of ecosystem management. Island Press. 1997, ISBN 978-1-55963-399-4
- Diamond, Jared (2005). Collapse: How Societies Choose to Fail Or Succeed. London: Penguin. p. 479. ISBN 0-14-303655-6.
- Eissing, Stefanie; Amend, Thora (2008). La protección de la naturaleza es divertida: manejo de áreas protegidas y comunicación ambiental : ideas procedentes de Panamá. Eschborn: GTZ. ISBN 978-3-925064-52-4.
- Gillis, Justin (2016-05-16). "In Latin America, Forests May Rise to Challenge of Carbon Dioxide". NYT. Retrieved 2016-05-17.
- "Philippines 5th out of 234 countries on forest gain". The Philippine Star. February 22, 2016. Retrieved February 22, 2016.

- "Explainer: back burning and fuel reduction". The Conversation. The Conversation Trust (UK) Limited. 8 August 2014. Retrieved 5 January 2016.
- "Forest Service Chief testifies before Senate appropriations committee on 2013 agency budget". US Forest Service. 18 April 2012. Retrieved 29 April 2012.
- "Sustainable Forest Management". Key Timberland Statistics. Weyerhaeuser. 10 June 2011. Retrieved 7 January 2012.
- Helliwell R., Wilson E. R. (2012). "Continuous cover forestry in Britain: challenges and opportunities". Quarterly Journal of Forestry. 106 (3): 214–224.
- "Reasons For Prescribed Fire In Forest Resource Management - A Guide for Prescribed Fire in Southern Forests". Bugwood.org. 2003-03-24. Retrieved 2011-06-25.
- "UK: The Role of Fire in the Ecology of Heathland in Southern Britain (IFFN No. 18 - January 1998)". Fire.uni-freiburg.de. Retrieved 2011-06-25.
- A changing climate of opinion?". The Economist. The Economist Newspaper Limited. 387: 93–96. 2008. Retrieved 2010-08-29.
- Adam, David (2009-02-18). "Fifth of world carbon emissions soaked up by extra forest growth, scientists find". The Guardian. London. Retrieved 2010-05-22.

4

Types of Forests

This chapter is a compilation of the various types of forests. Some of the forests that have been explained in the following section are taiga, temperate forests and tropical rainforest. The difference between temperate forests and rainforests is the difference between their climates and the animals that adapt in these forests. The section helps the readers in understanding all the types of forests and the differences between them.

Taiga

Taiga also known as boreal forest or snow forest, is a biome characterized by coniferous forests consisting mostly of pines, spruces and larches.

The taiga is the world's largest biome apart from the oceans. In North America it covers most of inland Canada and Alaska as well as parts of the extreme northern continental United States (northern Minnesota through the Upper Peninsula of Michigan to Upstate New York and northern New England), where it is known as the Northwoods. In Eurasia, it covers most of Sweden, Finland, much of Norway, some lowland/coastal areas of Iceland, much of Russia from Karelia in the west to the Pacific Ocean (including much of Siberia), and areas of northern Kazakhstan, northern Mongolia, and northern Japan (on the island of Hokkaidō). However, the main tree species, the length of the growing season and summer temperatures vary. For example, the taiga of North America mostly consists of spruces; Scandinavian and Finnish taiga consists of a mix of spruce, pines and birch; Russian taiga has spruces, pines and larches depending on the region, while the Eastern Siberian taiga is a vast larch forest.

White spruce taiga, Denali Highway, Alaska Range, Alaska

A different use of the term taiga is often encountered in the English language, with "boreal forest" used in the United States and Canada to refer to only the more southerly part of the biome, while "taiga" is used to describe the more barren areas of the northernmost part of the biome approaching the tree line and the tundra biome. Hoffman (1958) discusses the origin of this differential use in North America and why it is an inappropriate differentiation of the Russian term. Although at high elevations taiga grades into alpine tundra through Krummholz, it is not an alpine biome only like subalpine forest, and much of taiga is lowlands.

Climate and Geography

The taiga in the river valley near Verkhoyansk, Russia, at 67°N, must deal with the coldest winter temperatures in the northern hemisphere, but the extreme continentality of the climate gives an average daily high of 22 °C (72 °F) in July.

Taiga is the world's largest land biome, and makes up 29% of the world's forest cover; the largest areas are located in Russia and Canada. The taiga is the terrestrial biome with the lowest annual average temperatures after the tundra and permanent ice caps. Extreme winter minimums in the northern taiga are typically lower than those of the tundra. The lowest reliably recorded temperatures in the Northern Hemisphere were recorded in the taiga of northeastern Russia. The taiga or boreal forest has a subarctic climate with very large temperature range between seasons, but the long and cold winter is the dominant feature. This climate is classified as *Dfc*, *Dwc*, *Dsc*, *Dfd* and *Dwd* in the Köppen climate classification scheme, meaning that the short summer (24-hr average 10 °C (50 °F) or more) lasts 1–3 months and always less than 4 months. In Siberian taiga the average temperature of the coldest month is between −6 °C (21 °F) and −50 °C (−58 °F). There are also some much smaller areas grading towards the oceanic *Cfc* climate with milder winters, whilst the extreme south and (in Eurasia) west of the taiga reaches into humid continental climates (*Dfb*, *Dwb*) with longer summers. The mean annual temperature generally varies from -5 °C to 5 °C, (23 °F to 41 °F) but there are taiga areas in eastern Siberia and interior Alaska-Yukon where the mean annual reaches down to -10 °C. (14 °F) According to some sources, the boreal forest grades into a temperate mixed forest when mean annual temperature reaches about 3 °C (37 °F).

Discontinuous permafrost is found in areas with mean annual temperature below 0 °C, whilst in the *Dfd* and *Dwd* climate zones continuous permafrost occurs and restricts growth to very shallow-rooted trees like Siberian larch. The winters, with average temperatures below freezing, last five to seven months. Temperatures vary from −54 °C to 30 °C (-65 °F to 86 °F) throughout the whole year. The summers, while short, are generally warm and humid. In much of the taiga, -20 °C (-4 °F) would be a typical winter day temperature and 18 °C (64 °F) an average summer day.

Boreal forest near Shovel Point in Tettegouche State Park, along the northern shore of Lake Superior in Minnesota.

The growing season, when the vegetation in the taiga comes alive, is usually slightly longer than the climatic definition of summer as the plants of the boreal biome have a lower threshold to trigger growth. In Canada, Scandinavia and Finland, the growing season is often estimated by using the period of the year when the 24-hour average temperature is +5 °C (41 °F) or more. For the Taiga Plains in Canada, growing season varies from 80 to 150 days, and in the Taiga Shield from 100 to 140 days. Some sources claim 130 days growing season as typical for the taiga. Other sources mention that 50–100 frost-free days are characteristic. Data for locations in southwest Yukon gives 80–120 frost-free days. The closed canopy boreal forest in Kenozersky National Park near Plesetsk, Arkhangelsk Province, Russia, on average has 108 frost-free days. The longest growing season is found in the smaller areas with oceanic influences; in coastal areas of Scandinavia and Finland, the growing season of the closed boreal forest can be 145–180 days. The shortest growing season is found at the northern taiga–tundra ecotone, where the northern taiga forest no longer can grow and the tundra dominates the landscape when the growing season is down to 50–70 days, and the 24-hr average of the warmest month of the year usually is 10 °C (50 °F) or less. High latitudes mean that the sun does not rise far above the horizon, and less solar energy is received than further south. But the high latitude also ensures very long summer days, as the sun stays above the horizon nearly 20 hours each day, with only around 6 hours of daylight occurring in the dark winters, depending on latitude. The areas of the taiga inside the Arctic Circle have midnight sun in mid-summer and polar night in mid-winter.

Lakes and other water bodies are common in the taiga. The Helvetinjärvi National Park, Finland, situated in the closed canopy taiga (mid-boreal to south-boreal) with mean annual temperature of 4 °C (39 °F).

The taiga experiences relatively low precipitation throughout the year (generally 200–750 mm annually, 1,000 mm in some areas), primarily as rain during the summer months, but also as fog and snow. This fog, especially predominant in low-lying areas during and after the thawing of frozen Arctic seas, means that sunshine is not abundant in the taiga even during the long summer days. As evaporation is consequently low for most of the year, precipitation exceeds evaporation, and is sufficient to sustain the dense vegetation growth. Snow may remain on the ground for as long as nine months in the northernmost extensions of the taiga ecozone.

In general, taiga grows to the south of the 10 °C July isotherm, but occasionally as far north as the 9 °C (48 °F) July isotherm. Rich in spruces, Scots pines in the western Siberian plain, the taiga is dominated by larch in Eastern Siberia, before returning to its original floristic richness on the Pacific shores. Two deciduous trees mingle throughout southern Siberia: birch and populus tremula.

Late September in the fjords near Narvik, Norway. This oceanic part of the forest can see more than 1,000 mm precipitation annually and has warmer winters than the vast inland taiga

The southern limit is more variable, depending on rainfall; taiga may be replaced by forest steppe south of the 15 °C (59 °F) July isotherm where rainfall is very low, but more typically extends south to the 18 °C (64 °F) July isotherm, and locally where rainfall is higher (notably in eastern Siberia and adjacent Outer Manchuria) south to the 20 °C (68 °F) July isotherm. In these warmer areas the taiga has higher species diversity, with more warmth-loving species such as Korean pine, Jezo spruce, and Manchurian fir, and merges gradually into mixed temperate forest or, more locally (on the Pacific Ocean coasts of North America and Asia), into coniferous temperate rainforests where oak and hornbeam appear and join the conifers, birch and populus tremula.

The area currently classified as taiga in Europe and North America (except Alaska) was recently glaciated. As the glaciers receded they left depressions in the topography that have since filled with water, creating lakes and bogs (especially muskeg soil) found throughout the taiga.

Yukon, Canada. Several of the world's longest rivers go through the taiga, including Ob, Yenisei, Lena, and Mackenzie.

In Sweden the taiga is associated with the Norrland terrain.

Soils

Taiga soil tends to be young and poor in nutrients. It lacks the deep, organically enriched profile present in temperate deciduous forests. The thinness of the soil is due largely to the cold, which hinders the development of soil and the ease with which plants can use its nutrients. Fallen leaves and moss can remain on the forest floor for a long time in the cool, moist climate, which limits their organic contribution to the soil; acids from evergreen needles further leach the soil, creating spodosol, also known as podzol. Since the soil is acidic due to the falling pine needles, the forest floor has only lichens and some mosses growing on it. In clearings in the forest and in areas with more boreal deciduous trees, there are more herbs and berries growing. Diversity of soil organisms in the boreal forest is high, comparable to the tropical rainforest.

Flora

Since North America and Asia used to be connected by the Bering land bridge, a number of animal and plant species (more animals than plants) were able to colonize both continents and are distributed throughout the taiga biome. Others differ regionally, typically with each genus having several distinct species, each occupying different regions of the taiga. Taigas also have some small-leaved deciduous trees like birch, alder, willow, and poplar; mostly in areas escaping the most extreme winter cold. However, the Dahurian larch tolerates the coldest winters in the Northern Hemisphere in eastern Siberia. The very southernmost parts of the taiga may have trees such as oak, maple, elm and lime scattered among the conifers, and there is usually a gradual transition into a temperate mixed forest, such as the eastern forest-boreal transition of eastern Canada. In the interior of the continents with the driest climate, the boreal forests might grade into temperate grassland.

Boreal forest near Lake Baikal in Russia

There are two major types of taiga. The southern part is the closed canopy forest, consisting of many closely spaced trees with mossy ground cover. In clearings in the forest, shrubs and wildflowers are common, such as the fireweed. The other type is the lichen woodland or sparse taiga, with trees that are farther-spaced and lichen ground cover; the latter is common in the northernmost taiga. In the northernmost taiga the forest cover is not only more sparse, but often stunted in growth form; moreover, ice pruned asymmetric black spruce (in North America) are often seen, with diminished foliage on the windward side. In Canada, Scandinavia and Finland, the boreal forest is usually divided into three subzones: The high boreal (north boreal) or taiga zone; the middle boreal (closed forest); and the southern boreal, a closed canopy boreal forest with some scattered temperate deciduous trees among the conifers, such as maple, elm and oak. This southern boreal forest experiences the longest and warmest growing season of the biome, and in some regions (including Scandinavia, Finland and western Russia) this subzone is commonly used for agricultural purposes. The boreal forest is home to many types of berries; some are confined to the southern and middle closed boreal forest (such as wild strawberry and partridgeberry); others grow in most areas of the taiga (such as cranberry and cloudberry), and some can grow in both the taiga and the low arctic (southern part of) tundra (such as bilberry, bunchberry and lingonberry).

The forests of the taiga are largely coniferous, dominated by larch, spruce, fir and pine. The woodland mix varies according to geography and climate so for example the Eastern Canadian forests ecoregion of the higher elevations of the Laurentian Mountains and the northern Appalachian Mountains in Canada is dominated by balsam fir *Abies balsamea*, while further north the Eastern Canadian Shield taiga of northern Quebec and Labrador is notably black spruce *Picea mariana* and tamarack larch *Larix laricina*.

Evergreen species in the taiga (spruce, fir, and pine) have a number of adaptations specifically for survival in harsh taiga winters, although larch, the most cold-tolerant of all trees, is deciduous. Taiga trees tend to have shallow roots to take advantage of the thin soils, while many of them seasonally alter their biochemistry to make them more resistant to freezing, called "hardening". The narrow conical shape of northern conifers, and their downward-drooping limbs, also help them shed snow.

Because the sun is low in the horizon for most of the year, it is difficult for plants to generate energy from photosynthesis. Pine, spruce and fir do not lose their leaves seasonally and are able to photosynthesize with their older leaves in late winter and spring when light is good but temperatures are still too low for new growth to commence. The adaptation of evergreen needles limits the water lost due to transpiration and their dark green color increases their absorption of sunlight. Although precipitation is not a limiting factor, the ground freezes during the winter months and plant roots are unable to absorb water, so desiccation can be a severe problem in late winter for evergreens.

Moss (*Ptilium crista-castrensis*) cover on the floor of taiga

Although the taiga is dominated by coniferous forests, some broadleaf trees also occur, notably birch, aspen, willow, and rowan. Many smaller herbaceous plants, such as ferns and occasionally ramps grow closer to the ground. Periodic stand-replacing wildfires (with return times of between 20–200 years) clear out the tree canopies, allowing sunlight to invigorate new growth on the forest floor. For some species, wildfires are a necessary part of the life cycle in the taiga; some, e.g. jack pine have cones which only open to release their seed after a fire, dispersing their seeds onto the newly cleared ground; certain species of fungi (such as morels) are also known to do this. Grasses grow wherever they can find a patch of sun, and mosses and lichens thrive on the damp ground and on the sides of tree trunks. In comparison with other biomes, however, the taiga has low biological diversity.

Jack pine cones and morels after fire in a boreal forest.

Coniferous trees are the dominant plants of the taiga biome. A very few species in four main genera are found: the evergreen spruce, fir and pine, and the deciduous larch. In North America, one or two species of fir and one or two species of spruce are dominant. Across Scandinavia and western Russia, the Scots pine is a common component of the taiga, while taiga of the Russian Far East and Mongolia is dominated by larch.

Fauna

Brown bear, Kamchatka peninsula. Brown bears are among the largest and most widespread taiga omnivores.

The boreal forest, or taiga, supports a relatively small range of animals due to the harshness of the climate. Canada's boreal forest includes 85 species of mammals, 130 species of fish, and an estimated 32,000 species of insects. Insects play a critical role as pollinators, decomposers, and as a part of the food web. Many nesting birds rely on them for food in the summer months. The cold winters and short summers make the taiga a challenging biome for reptiles and amphibians, which depend on environmental conditions to regulate their body temperatures, and there are only a few species in the boreal forest including red-sided garter snake, common European adder, blue-spotted salamander, northern two-lined salamander, Siberian salamander, wood frog, northern leopard frog, boreal chorus frog, American toad, and Canadian toad. Most hibernate underground in winter. Fish of the taiga must be able to withstand cold water conditions and be able to adapt to life under ice-covered water. Species in the taiga include

Alaska blackfish, northern pike, walleye, longnose sucker, white sucker, various species of cisco, lake whitefish, round whitefish, pygmy whitefish, Arctic lamprey, various grayling species, brook trout (including sea-run brook trout in the Hudson Bay area), chum salmon, Siberian taimen, lenok and lake chub.

The taiga is home to a number of large herbivorous mammals, such as moose and reindeer/caribou. Some areas of the more southern closed boreal forest also have populations of other deer species such as the elk (wapiti) and roe deer. The largest animal in the taiga is the wood bison, found in northern Canada, Alaska and has been newly introduced into the Russian far-east. Small mammals of the Taiga biome include rodent species including beaver, squirrel, North American porcupine and vole, as well as a small number of lagomorph species such as snowshoe hare and mountain hare. These species have adapted to survive the harsh winters in their native ranges. Some larger mammals, such as bears, eat heartily during the summer in order to gain weight, and then go into hibernation during the winter. Other animals have adapted layers of fur or feathers to insulate them from the cold. Predatory mammals of the taiga must be adapted to travel long distances in search of scattered prey or be able to supplement their diet with vegetation or other forms of food (such as raccoons). Mammalian predators of the taiga include Canada lynx, Eurasian lynx, stoat, Siberian weasel, least weasel, sable, American marten, North American river otter, European otter, American mink, wolverine, Asian badger, fisher, gray wolf, coyote, red fox, brown bear, American black bear, Asiatic black bear, polar bear (only small areas at the taiga - tundra ecotone) and Siberian tiger.

More than 300 species of birds have their nesting grounds in the taiga. Siberian thrush, white-throated sparrow, and black-throated green warbler migrate to this habitat to take advantage of the long summer days and abundance of insects found around the numerous bogs and lakes. Of the 300 species of birds that summer in the taiga only 30 stay for the winter. These are either carrion-feeding or large raptors that can take live mammal prey, including golden eagle, rough-legged buzzard (also known as the rough-legged hawk), and raven, or else seed-eating birds, including several species of grouse and crossbills.

Fire

Fire has been one of the most important factors shaping the composition and development of boreal forest stands (Rowe 1955); it is the dominant stand-renewing disturbance through much of the Canadian boreal forest (Amiro et al. 2001). The fire history that characterizes an ecosystem is its *fire regime*, which has 3 elements: (1) fire type and intensity (e.g., crown fires, severe surface fires, and light surface fires), (2) size of typical fires of significance, and (3) frequency or return intervals for specific land units (Heinselman 1981). The average time within a fire regime to burn an area equivalent to the total area of an ecosystem is its *fire rotation* (Heinselman 1973) or *fire cycle* (Van Wagner 1978). However, as Heinselman (1981) noted, each physiographic site tends to have its own return interval, so that some areas are skipped for long periods, while others might burn two-times or more often during a nominal fire rotation.

The dominant fire regime in the boreal forest is high-intensity crown fires or severe surface fires of very large size, often more than 10,000 ha, and sometimes more than 400,000 ha (Heinselman 1981). Such fires kill entire stands. Fire rotations in the drier regions of western Canada and Alaska average 50–100 years, shorter than in the moister climates of eastern Canada, where they may average 200 years or more. Fire cycles also tend to be long near the tree line in the subarctic spruce-lichen woodlands. The longest cycles, possibly 300 years, probably occur in the western boreal in floodplain white spruce (Heinselman 1981).

Amiro et al. (2001) calculated the mean fire cycle for the period 1980 to 1999 in the Canadian boreal forest (including taiga) at 126 years. Increased fire activity has been predicted for western Canada, but parts of eastern Canada may experience less fire in future because of greater precipitation in a warmer climate (Flannigan et al. 1998).

The mature boreal forest pattern in the south shows balsam fir dominant on well-drained sites in eastern Canada changing centrally and westward to a prominence of white spruce, with black spruce and tamarack forming the forests on peats, and with jack pine usually present on dry sites except in the extreme east, where it is absent (Rowe and Scotter 1973). The effects of fires are inextricably woven into the patterns of vegetation on the landscape, which in the east favour black spruce, paper birch, and jack pine over balsam fir, and in the west give the advantage to aspen, jack pine, black spruce, and birch over white spruce. Many investigators have reported the ubiquity of charcoal under the forest floor and in the upper soil profile, e.g., La Roi (1967). Charcoal in soils provided Bryson et al. (1965) with clues about the forest history of an area 280 km north of the then current tree line at Ennadai Lake, District Keewatin, Northwest Territories.

Two lines of evidence support the thesis that fire has always been an integral factor in the boreal forest: (1) direct, eye-witness accounts and forest-fire statistics, and (2) indirect, circumstantial evidence based on the effects of fire, as well as on persisting indicators (Rowe and Scotter 1973). The patchwork mosaic of forest stands in the boreal forest, typically with abrupt, irregular boundaries circumscribing homogenous stands, is indirect but compelling testimony to the role of fire in shaping the forest. The fact is that most boreal forest stands are less than 100 years old, and only in the rather few areas that have escaped burning are there stands of white spruce older than 250 years (Rowe and Scotter 1973). The prevalence of fire-adaptive morphologic and reproductive characteristics of many boreal plant species is further evidence pointing to a long and intimate association with fire. Seven of the ten commonest trees in the boreal forest—jack pine, lodgepole pine, aspen, balsam poplar (*Populus balsamifera*), paper birch, tamarack, black spruce—can be classed as pioneers in their adaptations for rapid invasion of open areas. White spruce shows some pioneering abilities, too, but is less able than black spruce and the pines to disperse seed at all seasons. Only balsam fir and alpine fir seem to be poorly adapted to reproduce after fire, as their cones disintegrate at maturity, leaving no seed in the crowns.

The oldest forests in the northwest boreal region, some older than 300 years, are of white spruce occurring as pure stands on moist floodplains (Rowe 1970). Here, the frequency of fire is much less than on adjacent uplands dominated by pine, black spruce and aspen. In contrast, in the Cordilleran region, fire is most frequent in the valley bottoms, decreasing upward, as shown by a mosaic of young pioneer pine and broadleaf stands below, and older spruce–fir on the slopes above (Rowe and Scotter 1973). Without fire, the boreal forest would become more and more homogeneous, with the long-lived white spruce gradually replacing pine, aspen, balsam poplar, and birch, and perhaps even black spruce, except on the peatlands (Raup and Denny 1950).

Threats

Human Activities

Plesetsk Cosmodrome is situated in the taiga

Large areas of Siberia's taiga have been harvested for lumber since the collapse of the Soviet Union. Previously, the forest was protected by the restrictions of the Soviet Forest Ministry, but with the collapse of the Union, the restrictions regarding trade with Western nations have vanished. Trees are easy to harvest and sell well, so loggers have begun harvesting Russian taiga evergreen trees for sale to nations previously forbidden by Soviet law.

In Canada, eight percent of the taiga is protected from development, the provincial government allows forest management to occur on Crown land under rigorous constraints.

The main forestry practice in the boreal forest of Canada is clearcutting, which involves cutting down most of the trees in a given area, then replanting the forest as a monocrop (one species of tree) the following season.

Some of the products from logged boreal forests include toilet paper, copy paper, newsprint, and lumber. More than 90% of boreal forest products from Canada are exported for consumption and processing in the United States. However, with the recession and fewer US homes being built, that has changed.

Some of the larger cities situated in this biome are Murmansk, Arkhangelsk, Yakutsk, Anchorage, Yellowknife, Tromsø, Luleå, and Oulu.

Most companies that harvest in Canadian forests are certified by an independent third party agency such as the Forest Stewardship Council (FSC), Sustainable Forests Initiative (SFI), or the Canadian Standards Association (CSA). While the certification process differs between these groups, all of them include forest stewardship, respect for aboriginal peoples, compliance with local, provincial or national environmental laws, forest worker safety, education and training, and other environmental, business, and social requirements. The prompt renewal of all harvest sites by planting or natural renewal is also required.

Climate Change

Seney National Wildlife Refuge

The zone of latitude occupied by the boreal forest has experienced some of the greatest temperature increases on Earth, especially during the last quarter of the twentieth century. Winter temperatures have increased more than summer temperatures. The number of days with extremely cold temperatures (e.g., −20 to −40 °C (-4 to -40 °F) has decreased irregularly but systematically in nearly all the boreal region, allowing better survival for tree-damaging insects. In summer, the daily low temperature has increased more than the daily high temperature. In Fairbanks, Alaska, the length of the frost-free season has increased from 60–90 days in the early twentieth century to about 120 days a century later. Summer warming has been shown to increase water stress and reduce tree growth in dry areas of the southern boreal forest in central Alaska, western Canada and portions of far eastern Russia. Precipitation is relatively abundant in Scandinavia, Finland, northwest Russia and eastern Canada, where a longer growth season (i.e. the period when sap flow is not impeded by frozen water) accelerate tree growth. As a consequence of this warming trend, the warmer parts of the boreal forests are susceptible to replacement by grassland, parkland or temperate forest.

In Siberia, the taiga is converting from predominantly needle-shedding larch trees to evergreen conifers in response to a warming climate. This is likely to further accelerate warming, as the evergreen trees will absorb more of the sun's rays. Given the vast size of the area, such a change has the potential to affect areas well outside of the region. In much of the boreal forest in Alaska, the growth of white spruce trees are stunted by unusually warm summers, while trees on some of the coldest fringes of the forest are experiencing faster growth than previously.

Lack of moisture in the warmer summers are also stressing the birch trees of central Alaska.

Insects

Recent years have seen outbreaks of insect pests in forest-destroying plagues: the spruce-bark beetle (*Dendroctonus rufipennis*) in Yukon and Alaska; the mountain pine beetle in British Columbia; the aspen-leaf miner; the larch sawfly; the spruce budworm (*Choristoneura fumiferana*); the spruce coneworm.

Pollution

The effect of sulphur dioxide on woody boreal forest species was investigated by Addison et al. (1984), who exposed plants growing on native soils and tailings to 15.2 $\mu mol/m^3$ (0.34 ppm) of SO_2 on CO_2 assimilation rate (NAR). The Canadian maximum acceptable limit for atmospheric SO_2 is 0.34 ppm. Fumigation with SO_2 significantly reduced NAR in all species and produced visible symptoms of injury in 2–20 days. The decrease in NAR of deciduous species (trembling aspen [*Populus tremuloides*], willow [*Salix*], green alder [*Alnus viridis*], and white birch [*Betula papyrifera*]) was significantly more rapid than of conifers (white spruce, black spruce [*Picea mariana*], and jack pine [*Pinus banksiana*]) or an evergreen angiosperm (Labrador tea) growing on a fertilized Brunisol. These metabolic and visible injury responses seemed to be related to the differences in S uptake owing in part to higher gas exchange rates for deciduous species than for conifers. Conifers growing in oil sands tailings responded to SO_2 with a significantly more rapid decrease in NAR compared with those growing in the Brunisol, perhaps because of predisposing toxic material in the tailings. However, sulphur uptake and visible symptom development did not differ between conifers growing on the 2 substrates.

Acidification of precipitation by anthropogenic, acid-forming emissions has been associated with damage to vegetation and reduced forest productivity, but 2-year-old white spruce that were subjected to simulated acid rain (at pH 4.6, 3.6, and 2.6) applied weekly for 7 weeks incurred no statistically significant (P 0.05) reduction in growth during the experiment compared with the background control (pH 5.6) (Abouguendia and Baschak 1987). However, symptoms of injury were observed in all treatments, the number of plants and the number of needles affected increased with increasing rain acidity and with time. Scherbatskoy and Klein (1983) found no significant effect of chlorophyll concentration in white spruce at pH 4.3 and 2.8, but Abouguendia and Baschak (1987) found a significant reduction in white spruce at pH 2.6, while the foliar sulphur content significantly greater at pH 2.6 than any of the other treatments.

Protection

Peat bog in Dalarna, Sweden. Bogs and peatland are widespread in the taiga. They are home to a unique flora, and store vast amounts of carbon. In western Eurasia, the Scots pine is common in the boreal forest.

Many nations are taking direct steps to protect the ecology of the taiga by prohibiting logging, mining, oil and gas production, and other forms of development. In February 2010 the Canadian government established protection for 13,000 square kilometres of boreal forest by creating a new 10,700-square-kilometre park reserve in the Mealy Mountains area of eastern Canada and a 3,000-square-kilometre waterway provincial park that follows alongside the Eagle River from headwaters to sea.

Two Canadian provincial governments, Ontario and Quebec, introduced measures in 2008 that would protect at least half of their northern boreal forest. Although both provinces admitted it will take years to plan, work with Aboriginal and local communities and ultimately map out precise boundaries of the areas off-limits to development, the measures are expected to create some of the largest protected areas networks in the world once completed. Both announcements came the following year after a letter signed by 1,500 scientists called on political leaders to protect at least half of the boreal forest.

The taiga stores enormous quantities of carbon, more than the world's temperate and tropical forests combined, much of it in wetlands and peatland. In fact, current estimates place boreal forests as storing twice as much carbon per unit area as tropical forests.

Natural Disturbance

One of the biggest areas of research and a topic still full of unsolved questions is the recurring disturbance of fire and the role it plays in propagating the lichen woodland. The phenomenon of wildfire by lightning strike is the primary determinant of understory vegetation and because of this, it is considered to be the predominant force behind community and ecosystem properties in the lichen woodland. The significance of fire is clearly evident when one considers that understory vegetation influences tree seedling germination in the short term and decomposition of biomass and nutrient availability in the long term. The recurrent cycle of large, damaging fire occurs approximately every 70 to 100 years. Understanding the dynamics of this ecosystem is entangled with discovering the successional paths that the vegetation exhibits after a fire. Trees, shrubs, and lichens all recover from fire-induced damage through vegetative reproduction as well as invasion by propagules. Seeds that have fallen and become buried provide little help in re-establishment of a species. The reappearance of lichens is reasoned to occur because of varying conditions and light/nutrient availability in each different microstate. Several different studies have been done that have led to the formation of the theory that post-fire development can be propagated by any of four pathways: self replacement, species-dominance relay, species replacement, or gap-phase self replacement. Self replacement is simply the re-establishment of the pre-fire dominant species. Species-dominance relay is a sequential attempt of tree species to establish dominance in the canopy. Species replacement is when fires occur in sufficient frequency to interrupt species dominance relay. Gap-Phase Self-Replacement is the least common and so far has only been documented in Western Canada. It is a self replacement of the surviving species into the canopy gaps after a fire kills another species. The particular pathway taken after a fire disturbance depends on how the landscape is able to support trees as well as fire frequency. Fire frequency has a large role in shaping the original inception of the lower forest line of the lichen woodland taiga.

It has been hypothesized by Serge Payette that the spruce-moss forest ecosystem was changed into the lichen woodland biome due to the initiation of two compounded strong disturbances: large fire and the appearance and attack of the spruce budworm. The spruce budworm is a deadly insect to the spruce populations in the southern regions of the taiga. J.P. Jasinski confirmed this theory five years later stating "Their [lichen woodlands] persistence, along with their previous moss forest histories and current occurrence adjacent to closed moss forests, indicate that they are an alternative stable state to the spruce–moss forests".

Temperate Forests

Temperate forests correspond to forest concentration formed in the northern and southern hemisphere, or in temperate regions. Main characteristics include: wide leaves, large and tall trees and non seasonal vegetation. Temperate forests can be further distinguished by weather patterns and geographical features that favor the predominance of certain kinds of trees. In temperate coniferous forests, evergreen conifers predominate, while in temperate broadleaf and mixed forests, a more even distribution exists between evergreen and deciduous trees. Temperate deciduous forests, a subgroup of temperate broadleaf forests, consist of trees that lose their leaves every year. Finally, temperate rainforests typically have heavy rainfall and dense humidity.

In the temperate region, winters are mild and rainfall is moderate.The forests have trees, shrubs and bushes.There is less undergrowth. Most of the trees are deciduous, which shed their leaves during the autumn. In some areas, the deciduous trees are replaced by coniferous trees. The main trees found in these forests are the great redwood, oak, ash, maple, birch, beech, poplar, elm and pine. In many areas, forests have been cleared to make way for cultivation. Hardwood evergreen trees which are widely spaced and are found in the Mediterranean region are olive, cork, oak, and stone pine. Temperate rain forests tend to get a high of 350cm of precipitation yearly, and a low of 200cm yearly on average.

Tannins of tropical woods tend to be of a catechin nature rather than of the gallic type present in temperate woods.

Temperate Deciduous Forest

Temperate forest in Germany just as the leaf canopy is opening

Temperate deciduous forests or temperate broad-leaf forests are dominated by trees that lose their leaves each year. They are found in areas with warm, moist summers and mild winters. The three major areas of this forest type occur in the Northern Hemisphere: eastern North America, East Asia, and Europe. Smaller areas occur in Australasia and southern South America. Examples of typical trees in the Northern Hemisphere's deciduous forests include oak, maple, beech, and elm,while in Southern Hemisphere, trees of the genus *Nothofagus* dominate this type of forest. The diversity of tree species is higher in regions where the winter is milder, and also in mountainous regions that provide an array of soil types and microclimates. One of the world's great protected examples of this forest type is found in Great Smoky Mountains National Park.

Ecology

Bloodroot, like many other spring ephemerals, flowers in the spring before the forest canopy appears

The principal factor operating in these forests is the seasonal appearance and disappearance of the canopy. Shade from the canopy limits the growth of many kinds of plants. Many species that are typical of these forests time their growth and flowering to the short period just before the canopy opens; hence they are known as spring ephemerals. Examples include trilliums and bloodroot. Most spring ephemerals are insect-pollinated, and the seeds themselves are often transported by ants, a mode of dispersal known as myrmecochory. There are a smaller number of species able to grow under the canopy, and even a few that grow during the period when leaves are being lost.

Many understory plants have leaf adaptions to cope with low light levels, and the need to exploit moving flecks of light on the forest floor. A few, such as Indian pipe and Corallorhiza orchids, have adapted to the shade by parasitism. The trees similarly are controlled by shade. Most tree seedlings require small gaps when trees fall and make space in order to regenerate. A few require larger gaps such as those produced by windstorms. Gradients of soil moisture, soil depth, elevation and aspect control the distribution of many trees, shrubs and herbaceous species. Some require unusual conditions such as steep slopes, infertile soil, and drought to escape competition from the more common tree species.

Many migratory birds time their arrival to coincide with the opening of the canopy, which provides the insects that are their principal food sources for raising young. The spring warblers of North America are a typical example; for example, the black-throated blue warbler. Owing to the availability of wood from standing and fallen trees, woodpeckers are frequently found. The pileated woodpecker is a typical large species. Fallen wood, known as coarse woody debris, provides shelter for many kinds of amphibians, particularly salamanders. Many well-known animals live in this kind of forest; a few examples include squirrels, which are an important canopy species, and bears, which hibernate in the winter. The top predators in deciduous forest were once wolves and cougars, along with species of weasel like the fisher.

Pileated woodpeckers depend upon dead or dying trees as a source of food and for constructing their nests.

Human Effects

Humans have often colonized areas in the temperate deciduous forest. They have harvested wood for timber and charcoal. During the settlement of North America, potash made from tree ashes were exported back to Europe as fertilizer. As a result, less than one quarter of original forests remain. Many forests are now small fragments dissected by fields and roads; these islands of green often differ substantially from the original forests, particularly along the edges. The introduction of exotic diseases continues to be a threat to forest trees, and hence the forest; examples include the loss of chestnut and elm. At the same time, species like deer, which are clearing rather than true forest animals, have expanded their range and proliferated in these altered landscapes. Large deer populations have deleterious effects on tree regeneration overall, but particularly for edible species including yew, yellow birch and hemlock. Deer grazing also has significant negative effects on the number and kind of herbaceous flowering plants. The continuing pressure to increase deer populations, and the continued killing of top carnivores, suggests that overgrazing by deer will continue to be a significant forest conservation problem. Objective criteria for the restoration of deciduous forest include large trees, coarse woody debris, spring ephemeral, and top predators.

Temperate Coniferous Forest

Temperate coniferous forest is a terrestrial biome found in temperate regions of the world with warm summers and cool winters and adequate rainfall to sustain a forest. In most temperate coniferous forests, evergreen conifers predominate, while some are a mix of conifers and broadleaf evergreen trees and/or broadleaf deciduous trees. Temperate evergreen forests are common in the coastal areas of regions that have mild winters and heavy rainfall, or inland in drier climates or mountain areas. Coniferous forests can be found in the United States, Canada, Europe, and Asia. Many species of tree inhabit these forests including cedar, cypress, Douglas fir, fir, juniper, pine, podocarpus, spruce, redwood and yew. The understory also contains a wide variety of herbaceous and shrub species.

A pine forest is an example of a temperate coniferous forest

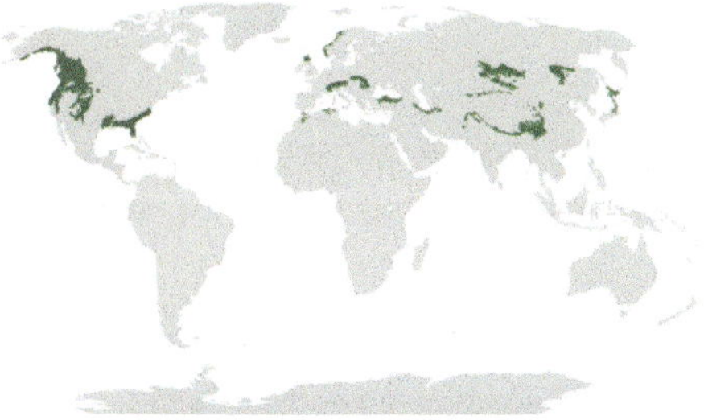

Map of temperate coniferous forest ecoregions of the world.

Structurally, these forests are rather simple, generally consisting of two layers: an overstory and understory. Some forests may support an intermediate layer of shrubs. Pine forests support a herbaceous understory that is generally dominated by grasses and herbaceous perennials, and are often subject to ecologically important wildfires.

Temperate rain forests occur only in seven regions around the world: the Pacific temperate rain forests of the Pacific Northwest, the Valdivian temperate rain forests of southwestern South America, the rain forests of New Zealand and Tasmania, northwest Europe (small pockets in Ireland, Scotland, Wales, Iceland and a somewhat larger area in Norway), southern Japan, and the eastern Black Sea-Caspian Sea region of Turkey and Georgia to northern Iran. The moist conditions of temperate rain forests generally support an understory of mosses, ferns and some shrubs. Temperate rain forests can be temperate coniferous forests or temperate broadleaf and mixed forests.

Species

The temperate coniferous rain forests around the world sustain the highest levels of biomass in any terrestrial ecosystem and are notable for trees of massive proportions, including coast redwood (*Sequoia sempervirens*), Douglas fir (*Pseudotsuga menzie-*

sii), Sitka spruce (*Picea sitchensis*), Alerce (*Fitzroya cupressoides*) and kauri (*Agathis australis*). These forests are quite rare, occurring in small areas of North America, southwestern South America and northern New Zealand. The Klamath-Siskiyou forests of northwestern California and southwestern Oregon is known for its rich variety of plant and animal species, including many endemic species.

The leaves on a coniferous tree are needle like. The leaves have a waxy coating to help the leaves from being damaged.

Temperate Coniferous Forest Ecoregions

Eurasia

Carpathian montane conifer forest, Slovakia.

Cedrus deodara in the Western Himalayan subalpine conifer forests.

North America

A temperate coniferous forest ecosystem in Petroglyphs Provincial Park, Ontario.

South America

Fitzroya forest at Alerce Costero National Park, Chile.

Bor

In Russia, a coniferous forest, specifically pine or spruce forest is known as bor (Russian: бор). Bors typically grow on well-drained sandy soils, sandy clay soils and mild loams. The names of several Russian historical buildings and localities refer to the word "bor" as the original place where they were built, such as Borovitskaya Tower or Church of the Savior on Bor.

Temperate Rainforest

Maples covered with epiphytic moss in the Hoh Rainforest

Temperate rainforests are coniferous or broadleaf forests that occur in the temperate zone and receive heavy rainfall.

Definition

Humid temperate rain forest in Termas Geométricas near Coñaripe, Chile

For temperate rain forests of North America, Alaback's definition is widely recognized:

Annual precipitation over 140 cm (55 in)

Mean annual temperature is between 4 and 12 °C (39 and 54 °F).

However, required annual precipitation depends on factors such as distribution of rainfall over the year, temperatures over the year and fog presence, and definitions in other countries differ considerably. For example, Australian definitions are ecological-structural rather than climatic:

Closed canopy of trees excludes at least 70% of the sky;

Forest is composed mainly of tree species which do not require fire for regeneration, but with seedlings able to regenerate under shade and in natural openings.

The latter would, for example, exclude a part of the temperate rain forests of western North America, as Coast Douglas-fir, one of its dominant tree species, requires stand-destroying disturbance to initiate a new cohort of seedlings. The North American definition would in turn exclude a part of temperate rain forests in other countries.

Global Distribution

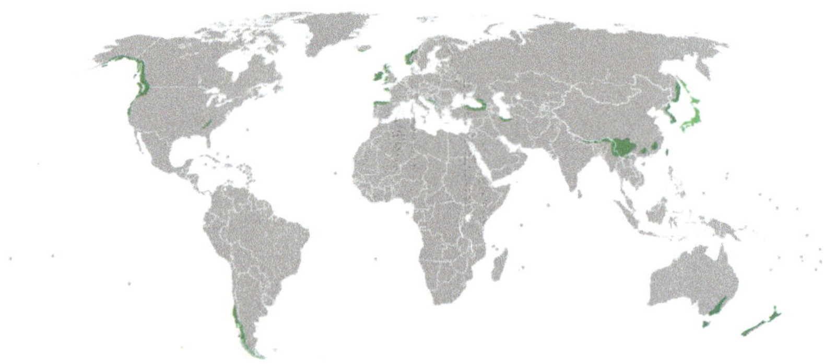

A map showing the areas of temperate rain forest

Temperate forests cover a large part of the Earth, but temperate rainforests only occur in a few regions around the world. Most of these occur in oceanic moist climates: the Pacific temperate rain forests in Western North America (Southeastern Alaska to Central California), the Valdivian and Magellanic temperate rainforests of southwestern South America (Southern Chile and adjacent Argentina), pockets of rain forest in Northwestern Europe (southern Norway to northern Iberia), temperate rainforests of southeastern Australia (Tasmania and Victoria) and the New Zealand temperate rainforests (South Island's west coast).

Others occur in subtropical moist climates: South Africa's Knysna-Amatole coastal forests, the Colchian rainforests of the eastern Black Sea region (Turkey and Georgia), the Caspian temperate rainforests of Iran and Azerbaijan, the mountain temperate rainforests along eastern Taiwan's Pacific coast, the eastern coast of the Korean Peninsula along the length of the Baekdu Mountain Range and in the area surrounding Mt. Jiri and across the peninsula's southern coastline, southwest Japan's Taiheiyo forests, coastal New South Wales and New Zealand's North Island.

Some areas, however, such as the Rocky Mountains of British Columbia (BC), northern Idaho and northwestern Montana, the Rocky Mountain Trench in BC and Montana, and the Russian Far East (Ussuri, Outer Manchuria, Sakhalin) in Asia have more continental climates but get enough precipitation in both rain and snow to harbor significant pockets of temperate rainforest.

Scattered small pockets of temperate rainforest also exist along the Appalachian Mountains from northern Georgia to New England. The mountainous coniferous forests of the Changbai Mountains bordering China and North Korea are also a good example, containing some of the richest high-elevation coniferous evergreen forests in East Asia.

Temperate Rain Forest Regions by Continent

North America

Pacific Temperate Rain Forests

Temperate rain forest in the Mount Hood Wilderness, Oregon, USA. This area, on the west side of the mountain, receives over 2.5 meters of rain per year.

A portion of the temperate rain forest region of North America, the largest area of temperate zone rain forests on the planet, is the Pacific temperate rain forests ecoregion which occur on west-facing coastal mountains along the Pacific coast of North America, from Kodiak Island in Alaska to northern California, and are part of the Nearctic ecozone. In the different system established by the Commission for Environmental Cooperation, this same general region is classed as the Pacific Maritime Ecozone by Environment Canada and as the Marine West Coast Forest and Northwestern Forested Mountains Level II ecoregions by the United States Environmental Protection Agency. In terms of the floristic province system used by botany, the bulk of the region is the Rocky Mountain Floristic Region but a small southern portion is part of the California Floristic Province.

Coast Redwood forest in Redwood National Park

Sub-ecoregions of the Pacific temperate rain forest ecoregion as defined by the WWF include the Northern Pacific coastal forests, Queen Charlotte Islands ecoregion, Vancouver Island ecoregion, British Columbia mainland coastal forests, Central Pacific coastal forests, Southern Cascades forests ecoregion, Klamath-Siskiyou coastal forests, and Northern California coastal forests ecoregions. They vary in their species composition, but are all predominantly coniferous, sometimes with an understory of broadleaved trees and shrubs. Most of the precipitation occurs in winter but summer fogs moisture is extracted by the trees and produces a fog drip keeping the forest moist. The Northern California coastal forests are home to the Coast Redwood (*Sequoia sempervirens*), the world's tallest tree. In the other ecoregions, Coast Douglas-fir (*Pseudotsuga menziesii* var. *menziesii*), Sitka Spruce (*Picea sitchensis*), Western Hemlock (*Tsuga heterophylla*) and Western redcedar (*Thuja plicata*) are the most important tree species. A common feature of Pacific temperate rain forests of North America is the Nurse log, a fallen tree which as it decays, provides ecological facilitation to seedlings. Trees such as the Coast Douglas-fir, Western Hemlock, Western Red Cedar, Pacific Yew, and Vine Maple are more closely related to coniferous and deciduous trees in the temperate forests of East Asia.

Temperate rain forest in Wells Gray Provincial Park (in the Cariboo Mountains) in British Columbia, Canada.

Some of the largest expanses of old growth are found in Olympic National Park, Mount

Rainier National Park, Tongass National Forest, Mount St. Helens National Monument, Redwood National Park, and throughout British Columbia (including British Columbia's Coastal Mountain Ranges), with the coastal Great Bear Rainforest containing the largest expanses of old growth temperate rainforest found in the world.

British Columbia's Rocky Mountains, Cariboo Mountains, Rocky Mountain Trench (east of Prince George) and the Columbia Mountains of Southeastern British Columbia (west of the Canadian Rocky Mountains that extend into parts of Idaho and Northwestern Montana in the USA), which include the Selkirk Mountains, Monashee Mountains, and the Purcell Mountains, have the largest stretch of interior temperate coniferous rain forests. These inland rainforests have more continental climate with a large proportion of the precipitation falling as snow. Being closer to the Rocky Mountains, there is more of a diverse mammalian fauna. Some of the best interior rain forests are found in Mount Revelstoke National Park and Glacier National Park (Canada) in the Columbia Mountains.

Appalachian Temperate Rain Forests

Temperate rain forest in the Great Smoky Mountains National Park, part of the Appalachian Mountains.

Temperate rain forests are located in the southern Appalachian Mountains where orographic precipitation causes weather systems coming from the west and from the Gulf of Mexico to drop more precipitation than in surrounding areas. The largest of these forest blocks are located in western North Carolina, northern Georgia, and far eastern Tennessee, largely in the Pisgah, Nantahala, and Chattahoochee National Forests and nearby Gorges State Park. In addition, small areas in the highest elevations of the Great Smoky Mountains also receive substantial rainfall, with Clingmans Dome, for example, collecting about 2000 mm of precipitation per year.

Red spruce and Fraser fir are dominant canopy trees in high mountain areas. In higher elevation (over 1,980 metres (6,500 ft)), Fraser fir is dominant, in middle elevation (1,675 to 1,890 metres (5,495 to 6,201 ft)) red spruce and Fraser fir grow together, and in lower elevation (1,370 to 1,650 metres (4,490 to 5,410 ft)) red spruce is dominant. Yellow birch, mountain ash, and mountain maple grow in the understory.Younger spruce and fir and shrubs like raspberry, blackberry, hobblebush, southern mountain cranberries, red elderberry, minniebush, southern bush honeysuckle are understory vegetation. Below the spruce-fir forest, at around 1,200 metres (3,900 ft), are forests of American beech, yellow birch, maple birch, and oak. Skunk cabbage and ground juniper are northern species that were pushed into the areas from the north.

The mild and wet environment supports the high diversity of fungi. Over 2000 species live in this area and scientists estimate many unidentified fungi may be there.

South America

Valdivian and Magellanic Temperate Rainforests

Aextoxicon punctatum forest in Punta Curiñanco

The temperate rain forests of South America are located on the Pacific coast of southern Chile, on the west-facing slopes of the southern Chilean coast range, and the Andes Mountains in both Chile and Argentina down to the southern tip of South America, and are part of the Neotropic ecozone. Temperate rain forests occur in the Valdivian temperate rain forests and Magellanic subpolar forests ecoregions. The Valdivian rainforests are home to a variety of broadleaf evergreen trees, like *Aextoxicon punctatum*, *Eucryphia cordifolia*, and southern beech (*Nothofagus*), but include many conifers as well, notably Alerce (*Fitzroya cupressoides*), one of the largest tree species of the world.

The Valdivian and Magellanic temperate rainforests are the only temperate rain forests in South America. Together they are the second largest in the world, after the Pacific temperate rain forests of North America. The Valdivian forests are a refuge for the Antarctic flora, and share many plant families and genera with the temperate rainforests of New Zealand, Tasmania, and Australia. Fully half the species of woody plants are endemic to this ecoregion.

In the Valdivian region the Andean Cordillera intercepts moist westerly winds along the Pacific coast during winter and summer months; these winds cool as they ascend the mountains, creating heavy rainfall on the mountains' west-facing slopes. The northward-flowing oceanic Humboldt Current creates humid and foggy conditions near the coast. The tree line is at about 2,400 m in the northern part of the ecoregion (35°S), and descends to 1,000 m in the south of the Valdivian region. In the summer the temperature can climb to 16.5 °C (61.7 °F), while during winter the temperature can drop below 7 °C (45 °F).

Africa

Knysna-amatole Coastal Rain Forests (South Africa)

Knysna Forest Biome near Nature's Valley, in the Tsitsikamma, South Africa

The temperate rainforests of South Africa are part of the Knysna-Amatole forests that are located along South Africa's Garden Route between Cape Town and Durban on the south-facing slopes of South Africa's Drakensberg Mountains facing the Indian Ocean. There are several coniferous podocarps that grow here. This forest receives a lot of moisture as fog from the Indian Ocean, and resembles not only other temperate rain forests worldwide, but also the montane evergreen Afromontane forests that occur at higher elevations in southern and eastern Africa. A fine example of this forest is in South Africa's Tsitsikamma National Park.

Europe

Temperate rainforest occurs in fragments across the north and west of Europe in countries such as southern Norway and northern Spain. Other temperate rainforest regions include areas of south eastern Europe such as mountains on the east coast of the Adriatic Sea, surrounding North Western Bulgaria along with the Black Sea.

Atlantic Oakwood Forest (United Kingdom and Ireland)

The woodlands are variously referred to in Britain as Upland Oakwoods, Atlantic Oakwoods, Western Oakwoods or Temperate Rainforest, Caledonian forest, and colloquially as 'Celtic Rainforests'. They are also listed in the British National Vegetation Classification as British NVC community W11 and British NVC community W17 depending on the ground flora. The majority of surviving fragments of Atlantic Oakwoods in Britain occur on steep-sided slopes above rivers and lakes which have avoided clearance and intensive grazing pressure. There are notable examples on the islands and shores of Loch Maree, Loch Sunart, Loch Lomond and one of the best preserved sites on the remote Taynish Peninsula in Argyll. There are also small areas on steep-sided riverine gorges in Snowdonia and Mid Wales. In England, they occur in the Lake District (Borrowdale Woods) and steep-sided riverine and estuarine valleys in Devon and Cornwall including the Fowey valley in Cornwall and the valley of the river Dart which flows off Dartmoor and has rainfall in excess of 2 metres per year.

Colchian (Colchis) Rain Forests (Turkey and Georgia)

The Colchian rainforests are found around the southeast corner of the Black Sea in Turkey and Georgia and are part of the Euxine-Colchic deciduous forests ecoregion, together with the drier Euxine forests further west. The Colchian rain forests are mixed, with deciduous black alder *(Alnus glutinosa)*, hornbeam (*Carpinus betulus* and *C. orientalis*), Oriental beech (*Fagus orientalis*), and sweet chestnut *(Castanea sativa)* together with evergreen Nordmann fir (*Abies nordmanniana*, the tallest tree in Europe at 78 m), Caucasian spruce (*Picea orientalis*) and Scots pine (*Pinus sylvestris*). The refugium is the largest throughout the Western Asian / near Eastern region. The area has multiple representatives of disjunct relict groups of plants with the closest relatives in Eastern Asia, southern Europe, and even North America. Over 70 species of forest snails of the region are endemic. Some relict species of vertebrates are Caucasian parsley frog, Caucasian salamander, Robert's snow vole and Caucasian grouse; they are almost entirely endemic groups of animals such as lizards of genus *Darevskia*. In general, species composition of this refugium is quite distinct and differs from that of the other Western Eurasian refugia. Genetic data suggest that the Colchis temperate rainforest, during the Ice Age, was fragmented into smaller parts; in particular, evolutionary lineages of the Caucasian Salamander from the central and south-western Colchis remained isolated from one another during the entire Ice Age.

Fragas do Eume (Spain)

The Fragas do Eume is a natural park situated in Galicia, north-western Spain. *Fraga* is a Galician word for "natural woodland", (old-growth forest) and the park is an example of a temperate rainforest in which oak (Quercus robur and Quercus pyrenaica) is the climax vegetation. The protected area extends along the valley of the river Eume within the Ferrolterra municipalities of Pontedeume, Cabanas, A Capela, Monfero and As Pontes de García Rodríguez. Some 500 people reside within the park. The monastery of Saint John of Caaveiro also lies within the park. The area was declared a natural park (a level of protection lower than national park) in 1997. It is one of six natural parks in

Galicia. The European Union has recognised the park as a Site of Community Importance. There are a number of species of ferns. Invertebrate species include the Kerry slug and it is an important site for amphibians.

Asia

Caspian Hyrcanian Forest (Iran and Azerbaijan)

The Caspian Hyrcanian mixed forests ecoregion in northern Iran contains a jungle in the form of a rain forest which stretches from the east in the Khorasan province to the west in the Ardebil province, covering the other provinces of Gilan, Mazandaran, and Golestan. The Elburz or Alborz mountain range is the highest mountain range in the Middle East which captures the moisture of the Caspian Sea to its north and forms subtropical and temperate rain forests in the northern part of Iran. The Iranians call this forest and region *Shomal* which means *north* in Persian.

In southeast Azerbaijan, this ecoregion includes the Lankaran Lowland and the Talysh Mountains, the latter being evenly divided with Iran to the south. They are deciduous forests containing tree species such as black alder (*Alnus glutinosa* subsp. *barbata*), hornbeam (*Carpinus betulus* and *C. orientalis*), Caucasian wingnut (*Pterocarya fraxinifolia*), chestnut-leaved oak (*Quercus castaneifolia*), Caucasian oak (*Quercus macranthera*), oriental beech (*Fagus orientalis*), Persian ironwood (*Parrotia persica*) and Persian silk tree (*Albizia julibrissin*).

The existing protected areas in Azerbaijan include:

- Gizil-Agach State Reserve – 88.4 square kilometres (34.1 sq mi)
- Hirkan National Park – 427.97 square kilometres (165.24 sq mi)
- Zuvand National Park – 15 square kilometres (5.8 sq mi)
- Girkan State Reserve – 3 square kilometres (1.2 sq mi)

High Elevation Mountain Rain Forests (Taiwan)

These forests are found in eastern Taiwan and Taiwan's Central Mountain Ranges, part of the Taiwan subtropical evergreen forest region covering the higher elevations. Most of the lower elevations are covered by subtropical broadleaf evergreen forests, dominated by Chinese Cryptocarya (*Cryptocarya chinensis*), *Castanopsis hystrix* and Japanese Blue Oak (*Quercus glauca*). Higher elevations give way to temperate forests with large stands of old growth Taiwan Cypress (*Chamaecyparis taiwanensis*), Camphor tree (*Cinnamomum camphora*), maple (*Acer* spp.), Chinese yew (*Taxus chinensis*), Taiwan Hemlock (*Tsuga chinensis*), and Taiwan Douglas-fir (*Pseudotsuga sinensis var. wilsoniana*). These higher elevation forests include also giant conifers Formosan Cypress (*Chamaecyparis formosensis*) and Taiwania (*Taiwania cryptomerioides*) Some fine examples of forests are found in Yushan (Jade Mountain) National Park and Alishan.

Baekdu Mountain Range (Taebaek and Sobaek Mountain Ranges) and South Sea forests (Korea)

The forests that cover the mountains and valleys of the Bakedu Mountain Range – from Mt. Baekdu, in the north, to Mt. Jiri, in the southwest, forming the spine of the Korean Peninsula – and the southern coast and islands of the peninsula – including Jeju Island – feature a wide variety of conifers and broadleaf trees. Much of these forests are protected in mountain and marine national forests, such as in Hallyeohaesang National Park, which encompasses 150.14 km2 of mountainous forests spread out over 69 uninhabited islands and 30 inhabited islands in Korea's South Sea that provide a home to 1,142 plant species, including major species such as red pine, black pine, common camellia, serrata oak, and cork oak, as well as rare species such as nadopungnan (sedirea japonica), daeheongnan (cymbidium nipponicum) and the Korean winter hazel. Major animals species, such as otters, small-eared cats, and badgers also call Hallyeohaesang National Park home, and overall there are 25 mammal species, 115 bird species, 16 reptile species, 1,566 insect species, and 24 freshwater fish species found among the forested, mountains islands.

Seoraksan National Park covers 398.539㎢ of mountainous forests near the eastern coast of the Korean Peninsula, and is a UNESCO designated Biosphere Preservation District. Over 2,000 animal species live in Seoraksan, including the Korean goral, musk deer, and there are also more than 1,400 rare plant species, such as the edelweiss.

Taiheiyo (Pacific) Rain Forests (Japan)

Southwestern Japan's Taiheiyo evergreen forests region covers much of Shikoku and Kyūshū Islands, and the Southern/Pacific Ocean-facing side of Honshu ("Taiheiyo" is the Pacific Ocean, in Japanese). Here the natural forests are mainly broadleaf evergreen in lower elevations and deciduous in higher elevations. The limit occurs at 500–1000 metres depending on latitude. The main tree species are members of beech family (Fagaceae). In lower altitudes these include evergreen oaks (*Quercus* spp.), Japanese Chinquapin (*Castanopsis cuspidata*) and Japanese Stone Oak (*Lithocarpus edulis*), and in higher altitudes Japanese Blue Beech (*Fagus japonica*) and Siebold's beech (*Fagus crenata*).

Jōmon Sugi, the largest specimen of Japanese Cedar (*Cryptomeria japonica*), on Yakushima, Japan

Some of the best preserved examples of forest are found in Kirishima-Yaku National Park on the Island of Yakushima off of Kyūshū in a very wet climate (the annual rainfall is 4,000 to 10,000 mm depending on altitude). Because of relatively infertile soils on granite, Yakushima's forests in higher elevations are dominated by a giant conifer species, Japanese Cedar (*Cryptomeria japonica*), rather than deciduous forests typical of the mainland. Other areas include Mount Kirishima near Kagoshima in southern Kyūshū. On Southern Honshū, there is a forest with the Nachi Falls located in Yoshino-Kumano National Park. This particular area of Honshū has been described as one of the rainiest spots in Japan.

Temperate Rainforests of the Russian Far East

Temperate rainforest in Great Otway National Park, Victoria, Australia.

Subtropical rainforest in Border Ranges National Park, New South Wales, Australia.

Oceania

Australian Temperate Rainforests

In Australia rainforests occur near the mainland east coast and in Tasmania. There are warm-temperate and cool-temperate rainforests. They are broadleaf evergreen forests with the exception of montane rainforests of Tasmania. Eucalypt forests are not classified as rainforests although some eucalypt forest types receive high annual rainfall (to over 2000 mm in Tasmania), and in the absence of fire they may develop to rainforest. If these widespread wet sclerophyll forests were considered rainforests, the total area of rainforest in Australia would be much larger.

Warm-temperate rainforest replaces subtropical rainforest on poorer soils or with increasing altitude and latitude in New South Wales and Victoria. Cool-temperate rainforests are widespread in Tasmania (Tasmanian temperate rain forests ecoregion) and they can be found scattered from the World Heritage listed Border Ranges National Park and Lamington National Park on the NSW/Queensland border to Otway Ranges, Strzelecki Ranges, Dandenong Ranges and Tarra Bulga in Victoria. In the northern NSW they are usually dominated by Antarctic Beech (*Nothofagus moorei*), in the southern NSW by Pinkwood (*Eucryphia moorei*) and Coachwood (*Ceratopetalum apetalum*) and in Victoria and Tasmania by Myrtle Beech (*Nothofagus cunninghamii*), Southern Sassafras (*Atherosperma moschatum*) and Mountain Ash (*Eucalyptus regnans*). The montane rainforests of Tasmania are dominated by Tasmanian endemic conifers (mainly *Athrotaxis* spp.). They are dominated by Ferns such as *Cyathea cooperi, Cyathea australis, Dicksonia antarctica, Cyathea cunninghamii* and *Cyathea leichhardtiana*.

Fiordland National Park near Te Anau, New Zealand

New Zealand Temperate Rainforests

The temperate rainforests of New Zealand occur on the western shore of the South Island and on the North Island. The forests are made up of coniferous podocarps and broadleaf evergreen trees. The podocarps are abundant at lower elevations, while southern beech (*Nothofagus*) can be found on higher slopes and in the cooler southernmost rainforests. Ecoregions include the Fiordland temperate forests and Westland temperate rainforests.

Temperate Broadleaf and Mixed Forest

Temperate broadleaf and mixed forest is a temperate climate terrestrial biome, with broadleaf tree ecoregions, and with conifer and broadleaf tree mixed coniferous forest ecoregions.

The term 'Temperate broadleaf and mixed forest' is used by the World Wildlife Fund (WWF) in global biogeography as one of the biome designations to organize ecoregions under.

Temperate broadleaf and mixed forest in Yunnan, southwest China.

Ecology

The typical structure of these forests includes four layers. The uppermost layer is the canopy composed of tall mature trees ranging from 30 to 61 m (100 to 200 ft) high. Below the canopy is the three-layered, shade-tolerant understory that is roughly 9 to 15 m (30 to 50 ft) shorter than the canopy. The top layer of the understory is the sub-canopy composed of smaller mature trees, saplings, and suppressed juvenile canopy layer trees awaiting an opening in the canopy. Below the sub-canopy is the shrub layer, composed of low growing woody plants. Typically the lowest growing (and most diverse) layer is the ground cover or herbaceous layer.

Trees

In the Northern hemisphere, characteristic dominant broadleaf trees in this biome include oaks (*Quercus* spp.), beeches (*Fagus* spp.), maples (*Acer* spp.), or birches (*Betula* spp.). The term "mixed forest" comes from the inclusion of coniferous trees as a canopy component of some of these forests. Typical coniferous trees include: Pines (*Pinus* spp.), firs (*Abies* spp.), and spruces (*Picea* spp.). In some areas of this biome the conifers may be a more important canopy species than the broadleaf species. In the Southern hemisphere, endemic genera such as *Nothofagus* and *Eucalyptus* occupy this biome.

Climate

Temperate broadleaf and mixed forests occur in areas with distinct warm and cool season, which give it a moderate annual average temperature—3 to 15.6 °C (37 to 60 °F). These forests occur in relatively warm and rainy climates, sometimes also with a distinct dry season. A dry season occurs in the winter in East Asia and in summer on the wet fringe of the Mediterranean climate zones. Other areas have a fairly even distribution of rainfall; annual rainfall is typically over 600 mm (24 in) and often over 1,500 mm (59 in). Temperatures are typically moderate except in parts of Asia such as Ussuriland where temperate forests can occur despite very harsh conditions with very cold winters.

Tropical Rainforest

An area of the Amazon rainforest in Brazil. The tropical rainforests of South America contain the largest diversity of species on earth.

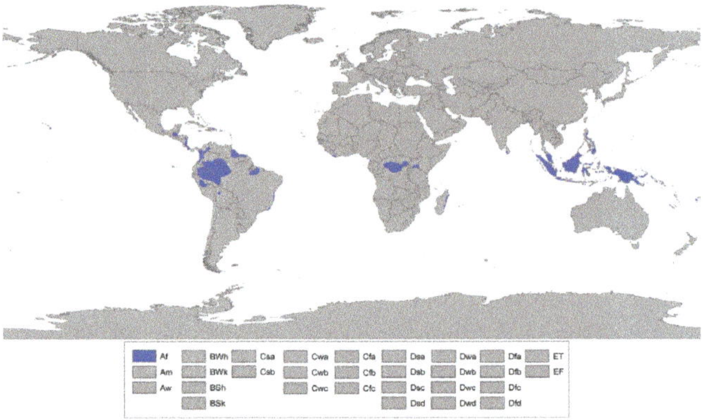

Tropical rainforest climate zones (Af).

Tropical rainforests occur in areas of tropical rainforest climate in which there is no dry season – all months have an average precipitation value of at least 60 mm – and may also be referred to as *lowland equatorial evergreen rainforest*. They rarely extend more than 10 degrees north or south of the equator are a sub-set of the tropical forest biome type that occurs roughly within the 28 degree latitudes (in the equatorial zone between the Tropic of Cancer and Tropic of Capricorn). Within the World Wildlife Fund's biome classification, tropical rainforests are a type of tropical moist broadleaf forest (or tropical wet forest) that also includes tropical seasonal forests.

Overview

Tropical rainforests can be characterized in two words: hot and wet. Mean monthly temperatures exceed 18 °C (64 °F) during all months of the year. Average annual rainfall is no less than 168 cm (66 in) and can exceed 1,000 cm (390 in) although it typically lies between 175 cm (69 in) and 200 cm (79 in). This high level of precipitation often results in poor soils due to leaching of soluble nutrients in the ground.

Amazon River rain forest in Peru

Tropical rainforests exhibit high levels of biodiversity. Around 40% to 75% of all biotic species are indigenous to the rainforests. Rainforests are home to half of all the living animal and plant species on the planet. Two-thirds of all flowering plants can be found in rainforests. A single hectare of rainforest may contain 42,000 different species of insect, up to 807 trees of 313 species and 1,500 species of higher plants. Tropical rainforests have been called the “world’s largest pharmacy”, because over one quarter of natural medicines have been discovered within them. It is likely that there may be many millions of species of plants, insects and microorganisms still undiscovered in tropical rainforests.

Tropical rainforests are among the most threatened ecosystems globally due to large-scale fragmentation as a result of human activity. Habitat fragmentation caused by geological processes such as volcanism and climate change occurred in the past, and have been identified as important drivers of speciation. However, fast human driven habitat destruction is suspected to be one of the major causes of species extinction. Tropical rain forests have been subjected to heavy logging and agricultural clearance throughout the 20th century, and the area covered by rainforests around the world is rapidly shrinking.

History

Tropical rainforests have existed on earth for hundreds of millions of years. Most tropical rainforests today are on fragments of the Mesozoic era supercontinent of Gondwana. The separation of the landmass resulted in a great loss of amphibian diversity while at the same time the drier climate spurred the diversification of reptiles. The division left tropical rainforests located in five major regions of the world: tropical America, Africa, Southeast Asia, Madagascar, and New Guinea, with smaller outliers in Australia. However, the specifics of the origin of rainforests remain uncertain due to an incomplete fossil record.

Types of Tropical Forest

Several biomes comprise the general term “tropical forest”:

Daintree rainforest in Queensland

Lowland Equatorial Evergreen Rainforests

Lowland equatorial evergreen rain forests are forests which receive high rainfall (more than 2000 mm, or 80 inches, annually) throughout the year. These forests occur in a belt around the equator, with the largest areas in the Amazon basin of South America, the Congo Basin of Central Africa, Philippines, Indonesia, and New Guinea.

Moist Tropical Seasonal Forest

Moist tropical seasonal forests receive high overall rainfall with a warm summer wet season and a cooler winter dry season. Some trees in these forests drop some or all of their leaves during the winter dry season, thus they are sometimes called "tropical mixed forest". They are found in parts of South America, in Central America and around the Caribbean, in coastal West Africa, parts of the Indian subcontinent, and across much of Indochina.

Montane Rainforests

Montane rainforests, some of which are known as cloud forests, are found in cooler-climate mountain areas. Depending on latitude, the lower limit of montane rainforests on large mountains is generally between 1500 and 2500 m while the upper limit is usually from 2400 to 3300 m.

Flooded Rainforests

Flooded forests are found in Amazon. Seven types of flooded forest are recognized for the Tambopata River Reserve in Amazonian Peru:

- Permanently waterlogged swamp forests are former oxbow lakes still flooded but covered in forest.
- Seasonally waterlogged swamp forests are oxbow lakes in the process of filling in.

- Lower floodplain forest are the lowest floodplain locations with a recognizable forest.
- Middle floodplain forests are tall occasionally flooded forests.
- Upper floodplain forests are tall rarely flooded forests.
- Old floodplain forests have been subjected to flooding within the last two hundred years.
- Previous floodplain are now *terra firme*, but were historically ancient floodplains.

Forest Structure

Rainforests are divided into different strata, or layers, with vegetation organized into a vertical pattern from the top of the soil to the canopy Each layer is a unique biotic community containing different plants and animals adapted for life in that particular strata. Only the emergent layer is unique to tropical rainforests, while the others are also found in temperate rainforests.

Forest Floor

Western lowland gorilla

The forest floor, the bottom-most layer, receives only 2% of the sunlight. Only plants adapted to low light can grow in this region. Away from riverbanks, swamps and clearings, where dense undergrowth is found, the forest floor is relatively clear of vegetation because of the low sunlight penetration. This more open quality permits the easy movement of larger animals such as: ungulates like the okapi (*Okapia johnstoni*), tapir (*Tapirus* sp.), Sumatran rhinoceros (*Dicerorhinus sumatrensis*), and apes like the western lowland gorilla (*Gorilla gorilla*), as well as many species of reptiles, amphibians, and insects. The forest floor also contains decaying plant and animal matter, which disappears quickly, because the warm, humid conditions promote rapid decay. Many forms of fungi growing here help decay the animal and plant waste.

Understory Layer

The understory layer lies between the canopy and the forest floor. The understory is home to a number of birds, small mammals, insects, reptiles, and predators. Examples include leopard (*Panthera pardus*), poison dart frogs (*Dendrobates* sp.), ring-tailed coati (*Nasua nasua*), boa constrictor (*Boa constrictor*), and many species of Coleoptera. The vegetation at this layer generally consists of shade-tolerant shrubs, herbs, small trees, and large woody vines which climb into the trees to capture sunlight. Only about 5% of sunlight breaches the canopy to arrive at the understory causing true understory plants to seldom grow to 3 m (10 feet). As an adaptation to these low light levels, understory plants have often evolved much larger leaves. Many seedlings that will grow to the canopy level are in the understory.

The canopy at the Forest Research Institute Malaysia

Canopy Layer

The canopy is the primary layer of the forest forming a roof over the two remaining layers. It contains the majority of the largest trees, typically 30–45 m in height. Tall, broad-leaved evergreen trees are the dominant plants. The densest areas of biodiversity are found in the forest canopy, as it often supports a rich flora of epiphytes, including orchids, bromeliads, mosses and lichens. These epiphytic plants attach to trunks and branches and obtain water and minerals from rain and debris that collects on the supporting plants. The fauna is similar to that found in the emergent layer, but more diverse. It is suggested that the total arthropod species richness of the tropical canopy might be as high as 20 million. Other species habituating this layer include many avian species such as the yellow-casqued wattled hornbill (*Ceratogymna elata*), collared sunbird (*Anthreptes collaris*), African gray parrot (*Psitacus erithacus*), keel-billed toucan (*Ramphastos sulfuratus*), scarlet macaw (*Ara macao*) as well as other animals like the spider monkey (*Ateles* sp.), African giant swallowtail (*Papilio antimachus*), three-toed sloth (*Bradypus tridactylus*), kinkajou (*Potos flavus*), and tamandua (*Tamandua tetradactyla*).

Emergent Layer

The emergent layer contains a small number of very large trees, called *emergents*, which grow above the general canopy, reaching heights of 45–55 m, although on occasion a few species will grow to 70–80 m tall. Some examples of emergents include: *Balizia elegans, Dipteryx panamensis, Hieronyma alchorneoides, Hymenolobium mesoamericanum, Lecythis ampla* and *Terminalia oblonga*. These trees need to be able to withstand the hot temperatures and strong winds that occur above the canopy in some areas. Several unique faunal species inhabit this layer such as the crowned eagle (*Stephanoaetus coronatus*), the king colobus (*Colobus polykomos*), and the large flying fox (*Pteropus vampyrus*).

However, stratification is not always clear. Rainforests are dynamic and many changes affect the structure of the forest. Emergent or canopy trees collapse, for example, causing gaps to form. Openings in the forest canopy are widely recognized as important for the establishment and growth of rainforest trees. It's estimated that perhaps 75% of the tree species at La Selva Biological Station, Costa Rica are dependent on canopy opening for seed germination or for growth beyond sapling size, for example.

Ecology

Climate

Artificial tropical rainforest in Barcelona

Tropical rainforests are located around and near the equator, therefore having what is called an equatorial climate characterized by three major climatic parameters: temperature, rainfall, and dry season intensity. Other parameters that affect tropical rainforests are carbon dioxide concentrations, solar radiation, and nitrogen availability. In general, climatic patterns consist of warm temperatures and high annual rainfall. However, the abundance of rainfall changes throughout the year creating distinct moist and dry seasons. Tropical forests are classified by the amount of rainfall received each year, which has allowed ecologists to define differences in these forests that look so similar in structure. According to Holdridge's classification of tropical ecosystems, true tropical rainforests have an annual rainfall greater than 800 cm and annual temperature greater than

24 degrees Celsius. However, most lowland tropical forests can be classified as tropical moist or wet forests, which differ in regards to rainfall. Tropical forest ecology- dynamics, composition, and function- are sensitive to changes in climate especially changes in rainfall. The climate of these forests is controlled by a band of clouds called the Intertropical Convergence Zone located near the equator and created by the convergence of the trade winds from the northern and Southern Hemispheres. The position of the band varies seasonally, moving north in the northern summer and south in the northern winter, and ultimately controlling the wet and dry seasons in the tropics. These regions have experienced strong warming at a mean rate of 0.26 degrees Celsius per decade which coincides with a global rise in temperature resulting from the anthropocentric inputs of greenhouse gases into the atmosphere. Studies have also found that precipitation has declined and tropical Asia has experienced an increase in dry season intensity whereas Amazonian has no significant pattern change in precipitation or dry season. Additionally, El Niño-Southern Oscillation events drive the inter-annual climatic variability in temperature and precipitation and result in drought and increased intensity of the dry season. As anthropogenic warming increases the intensity and frequency of ENSO will increase, rendering tropical rainforest regions susceptible to stress and increased mortality of trees and plants .

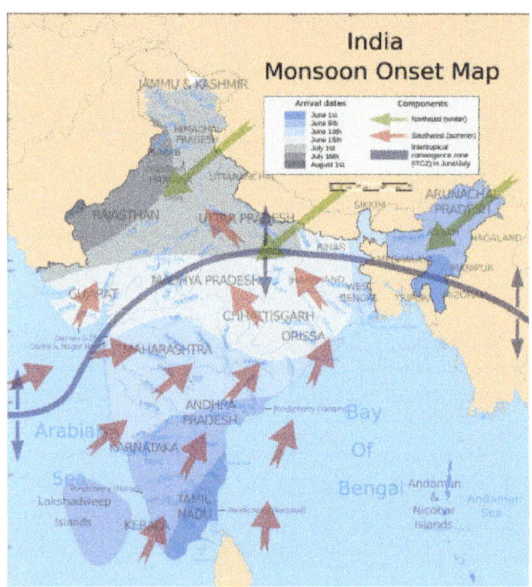

Onset dates and prevailing wind currents of the southwest summer monsoon.

Soils

Soil Types

Soil types are highly variable in the tropics and are the result of a combination of several variables such as climate, vegetation, topographic position, parent material, and soil age. Most tropical soils are characterized by significant leaching and poor nutrients, however there are some areas that contain fertile soils. Soils throughout the tropical rainforests fall into two classifications which include the ultisols and oxisols. Ultisols are known as well weathered, acidic red clay soils, deficient in major nutrients such as calcium and potassium. Similarly, oxisols are acidic, old, typically reddish, highly weathered and leached, however are well drained compared to ultisols. The clay content of ultisols is high, making it difficult for water to penetrate and flow through. The reddish color of both soils is the result of heavy heat and moisture forming oxides of iron and aluminium, which are insoluble in water and not taken up readily by plants.

Soil chemical and physical characteristics are strongly related to above ground productivity and forest structure and dynamics. The physical properties of soil control the tree turnover rates whereas chemical properties such as available nitrogen and phosphorus control forest growth rates. The soils of the eastern and central Amazon as well as the Southeast Asian Rainforest are old and mineral poor whereas the soils of the western Amazon (Ecuador and Peru) and volcanic areas of Costa Rica are young and mineral rich. Primary productivity or wood production is highest in western Amazon and lowest in eastern Amazon which contains heavily weathered soils classified as oxisols. Additionally, Amazonian soils are greatly weathered, making them devoid of minerals like phosphorus, potassium, calcium, and magnesium, which come from rock sources. However, not all tropical rainforests occur on nutrient poor soils, but on nutrient rich floodplains and volcanic soils located in the Andean foothills, and volcanic areas of Southeast Asia, Africa, and Central America.

Oxisols, infertile, deeply weathered and severely leached, have developed on the ancient Gondwanan shields. Rapid bacterial decay prevents the accumulation of humus. The concentration of iron and aluminium oxides by the laterization process gives the oxisols a bright red color and sometimes produces minable deposits (e.g., bauxite). On younger substrates, especially of volcanic origin, tropical soils may be quite fertile.

Nutrient Recycling

This high rate of decomposition is the result of phosphorus levels in the soils, precipitation, high temperatures and the extensive microorganism communities. In addition to the bacteria and other microorganisms, there are an abundance of other decomposers such as fungi and termites that aid in the process as well. Nutrient recycling is important because below ground resource availability controls the above ground biomass and community structure of tropical rainforests. These soils are typically phosphorus limited, which inhibits net primary productivity or the uptake of carbon. The soil contains microbial organisms such as bacteria, which break down leaf litter and other organic matter into inorganic forms of carbon usable by plants through a process called decomposition. During the decomposition process the microbial community is respiring, taking up oxygen and releasing carbon dioxide. The decomposition rate can be evaluated by measuring the uptake of oxygen. High temperatures and precipitation increase decomposition rate, which allows plant litter to rapidly decay in tropical regions, releasing nutrients that are immediately taken up by plants through surface or ground waters. The seasonal patterns in respiration are controlled by leaf litter fall and precipitation, the driving force moving the decomposable carbon from the litter to the soil. Respiration rates are highest early in the wet season because the recent dry season results in a large percentage of leaf litter and thus a higher percentage of organic matter being leached into the soil.

Buttress Roots

A common feature of many tropical rainforests is the distinct buttress roots of trees. Instead of penetrating to deeper soil layers, buttress roots create a widespread root network at the surface for more efficient uptake of nutrients in a very nutrient poor and competitive environment. Most of the nutrients within the soil of a tropical rainforest occur near the surface because of the rapid turnover time and decomposition of organisms and leaves. Because of this, the buttress roots occur at the surface so the trees can maximize uptake and actively compete with the rapid uptake of other trees. These roots also aid in water uptake and storage, increase surface area for gas exchange, and collect leaf litter for added nutrition. Additionally, these roots reduce soil erosion and maximize nutrient acquisition during heavy rains by diverting nutrient rich water flowing down the trunk into several smaller flows while also acting as a barrier to ground flow. Also, the large surface areas these roots create provide support and stability to rainforests trees, which commonly grow to significant heights. This added stability allows these trees to withstand the impacts of severe storms, thus reducing the occurrence of fallen trees.

Forest Succession

Succession is an ecological process that changes the biotic community structure over time towards a more stable, diverse community structure after an initial disturbance to the community. The initial disturbance is often a natural phenomenon or human caused event. Natural disturbances include hurricanes, volcanic eruptions, river movements or an event as small as a fallen tree that creates gaps in the forest. In tropical rainforests, these same natural disturbances have been well documented in the fossil record, and are credited with encouraging speciation and endemism.

Biodiversity and Speciation

Young orangutan at Bukit Lawang, Sumatra

Tropical rainforests exhibit a vast diversity in plant and animal species. The root for this remarkable speciation has been a query of scientists and ecologists for years. A number of theories have been developed for why and how the tropics can be so diverse.

Interspecific Competition

Interspecific competition results from a high density of species with similar niches in the tropics and limited resources available. Species which "lose" the competition may either become extinct or find a new niche. Direct competition will often lead to one species dominating another by some advantage, ultimately driving it to extinction. Niche partitioning is the other option for a species. This is the separation and rationing of necessary resources by utilizing different habitats, food sources, cover or general behavioral differences. A species with similar food items but different feeding times is an example of niche partitioning.

Pleistocene Refugia

The theory of Pleistocene refugia was developed by Jürgen Haffer in 1969 with his article *Speciation of Amazonian Forest Birds*. Haffer proposed the explanation for speciation was the product of rainforest patches being separated by stretches of non forest vegetation during the last glacial period. He called these patches of rainforest areas refuges and within these patches allopatric speciation occurred. With the end of the glacial period and increase in atmospheric humidity, rainforest began to expand and the refuges reconnected. This theory has been the subject of debate. Scientists are still skeptical of whether or not this theory is legitimate. Genetic evidence suggests speciation had occurred in certain taxa 1–2 million years ago, preceding the Pleistocene.

Human Dimensions

Habitation

Tropical rainforests have harboured human life for many millennia, with many Indian tribes in South- and Central America, who belong to the Indigenous peoples of the Americas, the Congo Pygmies in Central Africa, and several tribes in South-East Asia, like the Dayak people and the Penan people in Borneo. Food resources within the forest are extremely dispersed due to the high biological diversity and what food does exist is largely restricted to the canopy and requires considerable energy to obtain. Some groups of hunter-gatherers have exploited rainforest on a seasonal basis but dwelt primarily in adjacent savanna and open forest environments where food is much more abundant. Other people described as rainforest dwellers are hunter-gatherers who subsist in large part by trading high value forest products such as hides, feathers, and honey with agricultural people living outside the forest.

Indigenous Peoples

Members of an uncontacted tribe encountered in the Brazilian state of Acre in 2009.

A variety of indigenous people live within the rainforest as hunter-gatherers, or subsist as part-time small scale farmers supplemented in large part by trading high-value forest products such as hides, feathers, and honey with agricultural people living outside the forest. Peoples have inhabited the rainforests for tens of thousands of years and have remained so elusive that only recently have some tribes been discovered. These indigenous peoples are greatly threatened by loggers in search for old-growth tropical hardwoods like Ipe, Cumaru and Wenge, and by farmers who are looking to expand their land, for cattle(meat), and soybeans, which are used to feed cattle in Europe and China. On 18 January 2007, FUNAI reported also that it had confirmed the presence of 67 different uncontacted tribes in Brazil, up from 40 in 2005. With this addition, Brazil has now overtaken the island of New Guinea as the country having the largest number of uncontacted tribes. The province of Irian Jaya or West Papua in the island of New Guinea is home to an estimated 44 uncontacted tribal groups.

The pygmy peoples are hunter-gatherer groups living in equatorial rainforests characterized by their short height (below one and a half meters, or 59 inches, on average). Amongst this group are the Efe, Aka, Twa, Baka, and Mbuti people of Central Africa. However, the term pygmy is considered pejorative so many tribes prefer not to be labeled as such.

Some notable indigenous peoples of the Americas, or Amerindians, include the Huaorani, Ya□nomamö, and Kayapo people of the Amazon. The traditional agricultural system practiced by tribes in the Amazon is based on swidden cultivation (also known as slash-and-burn or shifting cultivation) and is considered a relatively benign disturbance. In fact, when looking at the level of individual swidden plots a number of traditional farming practices are considered beneficial. For example, the use of shade trees and fallowing all help preserve soil organic matter, which is a critical factor in the maintenance of soil fertility in the deeply weathered and leached soils common in the Amazon.

There is a diversity of forest people in Asia, including the Lumad peoples of the Philippines and the Penan and Dayak people of Borneo. The Dayaks are a particularly interesting group as they are noted for their traditional headhunting culture. Fresh human heads were required to perform certain rituals such as the Iban "kenyalang" and the Kenyah "mamat". Pygmies who live in Southeast Asia are, amongst others, referred to as "Negrito".

Resources

Cultivated Foods and Spices

Yam, coffee, chocolate, banana, mango, papaya, macadamia, avocado, and sugarcane all originally came from tropical rainforest and are still mostly grown on plantations in regions that were formerly primary forest. In the mid-1980s and 1990s, 40 million tons of bananas were consumed worldwide each year, along with 13 million tons of mango. Central American coffee exports were worth US$3 billion in 1970. Much of the genetic variation used in evading the damage caused by new pests is still derived from resistant wild stock. Tropical forests have supplied 250 cultivated kinds of fruit, compared to only 20 for temperate forests. Forests in New Guinea alone contain 251 tree species with edible fruits, of which only 43 had been established as cultivated crops by 1985.

Ecosystem services

In addition to extractive human uses, rain forests also have non-extractive uses that are frequently summarized as ecosystem services. Rain forests play an important role in maintaining biological diversity, sequestering and storing carbon, global climate regulation, disease control, and pollination.

Tourism

Canopy walkway for seeing the diverse tropical forest in Costa Rica

Despite the negative effects of tourism in the tropical rainforests, there are also several important positive effects.

- In recent years ecotourism in the tropics has increased. While rainforests are becoming increasingly rare, people are travelling to nations that still have this diverse habitat. Locals are benefiting from the additional income brought in by visitors, as well areas deemed interesting for visitors are often conserved. *Ecotourism can be an incentive for conservation, especially when it triggers positive economic change.* Ecotourism can include a variety of activities including animal viewing, scenic jungle tours and even viewing cultural sights and native villages. If these practices are performed appropriately this can be beneficial for both locals and the present flora and fauna.

- An increase in tourism has increased economic support, allowing more revenue to go into the protection of the habitat. Tourism can contribute directly to the conservation of sensitive areas and habitat. Revenue from park-entrance fees and similar sources can be utilised specifically to pay for the protection and management of environmentally sensitive areas. Revenue from taxation and tourism provides an additional incentive for governments to contribute revenue to the protection of the forest.

- Tourism also has the potential to increase public appreciation of the environment and to spread awareness of environmental problems when it brings people into closer contact with the environment. Such increased awareness can induce more environmentally conscious behavior. Tourism has had a positive effect on wildlife preservation and protection efforts, notably in Africa but also in South America, Asia, Australia, and the South Pacific.

Conservation

Threats

Mining and Drilling

The Ok Tedi Mine in southwestern Papua New Guinea

Deposits of precious metals (gold, silver, coltan) and fossil fuels (oil and natural gas) occur underneath rainforests globally. These resources are important to developing nations and their extraction is often given priority to encourage economic growth. Mining and drilling can require large amounts of land development, directly causing deforestation. In Ghana, a West African nation, deforestation from decades of mining activity left about 12% of the country's original rainforest intact.

Conversion to Agricultural Land

With the invention of agriculture, humans were able to clear sections of rainforest to produce crops, converting it to open farmland. Such people, however, obtain their food primarily from farm plots cleared from the forest and hunt and forage within the forest to supplement this. The issue arising is between the independent farmer providing for his family and the needs and wants of the globe as a whole. This issue has seen little improvement because no plan has been established for all parties to be aided.

Agriculture on formerly forested land is not without difficulties. Rainforest soils are often thin and leached of many minerals, and the heavy rainfall can quickly leach nutrients from area cleared for cultivation. People such as the Yanomamo of the Amazon, utilize slash-and-burn agriculture to overcome these limitations and enable them to push deep into what were previously rainforest environments. However, these are not rainforest dwellers, rather they are dwellers in cleared farmland that make forays into the rainforest. Up to 90% of the typical Yanamomo diet comes from farmed plants.

Some action has been taken by suggesting fallow periods of the land allowing secondary forest to grow and replenish the soil. Beneficial practices like soil restoration and conservation can benefit the small farmer and allow better production on smaller parcels of land.

Climate Change

The tropics take a major role in reducing atmospheric carbon dioxide. The tropics (most notably the Amazon rainforest) are called carbon sinks. As major carbon reducers and carbon and soil methane storages, their destruction contributes to increasing global energy trapping, atmospheric gases. Climate change has been significantly contributed to by the destruction of the rainforests. A simulation was performed in which all rainforest in Africa were removed. The simulation showed an increase in atmospheric temperature by 2.5 to 5 degrees Celsius.

Protection

Efforts to protect and conserve tropical rainforest habitats are diverse and widespread. Tropical rainforest conservation ranges from strict preservation of habitat to finding sustainable management techniques for people living in tropical rainforests. International policy has also introduced a market incentive program called Reducing Emissions from Deforestation and Forest Degradation (REDD) for companies and governments to outset their carbon emissions through financial investments into rainforest conservation.

References

- Floyd, A. (1990) Australian Rainforests in New South Wales, Vol. 1. Surrey Beatty & Sons Pty Ltd, Chipping Norton, NSW, ISBN 0949324302.
- Franklin, J.F. & Dyrness C.T. (1988) Natural Vegetation of Oregon and Washington. Oregon State University Press, ISBN 0870713566.
- van Zeist W, Bottema S. (1991). Late Quaternary vegetation of the Near East. Weisbaden: Reichert, ISBN 3882265302.
- Farjon, A. (2005). Monograph of Cupressaceae and Sciadopitys. Royal Botanic Gardens, Kew. ISBN 978-1-84246-068-9
- Harden, G., McDonald, B. & Williams, J. (2006). Rainforest Trees and Shrubs. Gwen Harden Publishing, Nambucca Heads. ISBN 978-0-9775553-0-7
- Newman, Arnold (2002). Tropical Rainforest: Our Most Valuable and Endangered Habitat With a Blueprint for Its Survival Into the Third Millennium (2 ed.). Checkmark. ISBN 0816039739.
- Why the Amazon Rainforest is So Rich in Species. Earthobservatory.nasa.gov (5 December 2005). Retrieved on 28 March 2013.
- Study reveals for first time true diversity of life in soils across the globe, new species discovered". Physorg.com. Retrieved 2012-01-14.
- Russian boreal forests undergoing vegetation change, study shows". Sciencedaily.com. 2011-03-25. doi:10.1111/j.1365-2486.2011.02417.x. Retrieved 2012-01-14.
- "Fairbanks Daily News-Miner - New study states boreal forests shifting as Alaska warms". News-miner.com. Retrieved 2012-01-14.
- TAMPERE/PIRKKALA, FINLAND Weather History and Climate Data". Worldclimate.com. 2007-02-04. Retrieved 2011-02-21.

5

Threats to Forests

Deforestation is one of the biggest threats to forests; deforestation is the clearance of forests. The purpose behind this clearance can be industrial or it can be done with the intention of converting these forests into farms. Alternatively, the threats of forests are forest dieback, illegal logging, clearcutting and wildfire. The aspects explained in this chapter are of vital importance, and provide a better understanding of forests.

Deforestation

Satellite photograph of deforestation in progress in the Tierras Bajas project in eastern Bolivia.

Deforestation, clearance or clearing is the removal of a forest or stand of trees where the land is thereafter converted to a non-forest use. Examples of deforestation include conversion of forestland to farms, ranches, or urban use. Tropical rainforests is where the most concentrated deforestation occurs. About 30% of Earth's land surface is covered by forests.

In temperate mesic climates, natural regeneration of forest stands often will not occur in the absence of disturbance, whether natural or anthropogenic. Furthermore, biodiversity after regeneration harvest often mimics that found after natural disturbance, including biodiversity loss after naturally occurring rainforest destruction.

Deforestation occurs for multiple reasons: trees are cut down to be used or sold as fuel (sometimes in the form of charcoal) or timber, while cleared land is used as pasture for livestock and plantation. The removal of trees without sufficient reforestation has resulted in damage to habitat, biodiversity loss and aridity. It has adverse impacts on biosequestration of atmospheric carbon dioxide. Deforestation has also been used in war to deprive the enemy of cover for its forces and also vital resources. Modern examples of this were the use of Agent Orange by the British military in Malaya during the Malayan Emergency and the United States military in Vietnam during the Vietnam War. As of 2005, net deforestation rates have ceased to increase in countries with a per capita GDP of at least US$4,600. Deforested regions typically incur significant adverse soil erosion and frequently degrade into wasteland.

Disregard of ascribed value, lax forest management and deficient environmental laws are some of the factors that allow deforestation to occur on a large scale. In many countries, deforestation, both naturally occurring and human induced, is an ongoing issue. Deforestation causes extinction, changes to climatic conditions, desertification, and displacement of populations as observed by current conditions and in the past through the fossil record. More than half of all plant and land animal species in the world live in tropical forests.

Between 2000 and 2012, 2.3 million square kilometres (890,000 square miles) of forests around the earth were cut down. As a result of deforestation, only 6.2 million square kilometres (2.4 million square miles) remain of the original 16 million square kilometres (6 million square miles) of forest that formerly covered the earth.

Causes

According to the United Nations Framework Convention on Climate Change (UNFCCC) secretariat, the overwhelming direct cause of deforestation is agriculture. Subsistence farming is responsible for 48% of deforestation; commercial agriculture is responsible for 32% of deforestation; logging is responsible for 14% of deforestation and fuel wood removals make up 5% of deforestation.

Experts do not agree on whether industrial logging is an important contributor to global deforestation. Some argue that poor people are more likely to clear forest because they have no alternatives, others that the poor lack the ability to pay for the materials and labour needed to clear forest. One study found that population increases due to high fertility rates were a primary driver of tropical deforestation in only 8% of cases.

Other causes of contemporary deforestation may include corruption of government institutions, the inequitable distribution of wealth and power, population growth and overpopulation, and urbanization. Globalization is often viewed as another root cause of deforestation, though there are cases in which the impacts of globalization (new flows of labor, capital, commodities, and ideas) have promoted localized forest recovery.

In 2000 the United Nations Food and Agriculture Organization (FAO) found that "the role of population dynamics in a local setting may vary from decisive to negligible," and that deforestation can result from "a combination of population pressure and stagnating economic, social and technological conditions."

The last batch of sawnwood from the peat forest in Indragiri Hulu, Sumatra, Indonesia. Deforestation for oil palm plantation.

The degradation of forest ecosystems has also been traced to economic incentives that make forest conversion appear more profitable than forest conservation. Many important forest functions have no markets, and hence, no economic value that is readily apparent to the forests' owners or the communities that rely on forests for their well-being. From the perspective of the developing world, the benefits of forest as carbon sinks or biodiversity reserves go primarily to richer developed nations and there is insufficient compensation for these services. Developing countries feel that some countries in the developed world, such as the United States of America, cut down their forests centuries ago and benefited greatly from this deforestation, and that it is hypocritical to deny developing countries the same opportunities: that the poor shouldn't have to bear the cost of preservation when the rich created the problem.

Some commentators have noted a shift in the drivers of deforestation over the past 30 years. Whereas deforestation was primarily driven by subsistence activities and government-sponsored development projects like transmigration in countries like Indonesia and colonization in Latin America, India, Java, and so on, during late 19th century and the earlier half of the 20th century. By the 1990s the majority of deforestation was caused by industrial factors, including extractive industries, large-scale cattle ranching, and extensive agriculture.

Environmental Problems

Atmospheric

Deforestation is ongoing and is shaping climate and geography.

Deforestation is a contributor to global warming, and is often cited as one of the major causes of the enhanced greenhouse effect. Tropical deforestation is responsible for approximately 20% of world greenhouse gas emissions. According to the Intergovernmental Panel on Climate Change deforestation, mainly in tropical areas, could account for up to one-third of total anthropogenic carbon dioxide emissions. But recent calculations suggest that carbon dioxide emissions from deforestation and forest degradation (excluding peatland emissions) contribute about 12% of total anthropogenic carbon dioxide emissions with a range from 6 to 17%. Deforestation causes carbon dioxide to linger in the atmosphere. As carbon dioxide accrues, it produces a layer in the atmosphere that traps radiation from the sun. The radiation converts to heat which causes global warming, which is better known as the greenhouse effect. Plants remove carbon in the form of carbon dioxide from the atmosphere during the process of photosynthesis, but release some carbon dioxide back into the atmosphere during normal respiration. Only when actively growing can a tree or forest remove carbon, by storing it in plant tissues. Both the decay and burning of wood releases much of this stored carbon back to the atmosphere. In order for forests to take up carbon, there must be a net accumulation of wood. One way is for the wood to be harvested and turned into long-lived products, with new young trees replacing them. Deforestation may also cause carbon stores held in soil to be released. Forests can be either sinks or sources depending upon environmental circumstances. Mature forests alternate between being net sinks and net sources of carbon dioxide.

Illegal slash and burn practice in Madagascar, 2010

In deforested areas, the land heats up faster and reaches a higher temperature, leading to localized upward motions that enhance the formation of clouds and ultimately produce more rainfall. However, according to the Geophysical Fluid Dynamics Laboratory, the models used to investigate remote responses to tropical deforestation showed a broad but mild temperature increase all through the tropical atmosphere. The model predicted <0.2 °C warming for upper air at 700 mb and 500 mb. However, the model shows no significant changes in other areas besides the Tropics. Though the model showed no significant changes to the climate in areas other than the Tropics, this may not be the case since the model has possible errors and the results are never absolutely definite.

Fires on Borneo and Sumatra, 2006. People use
slash-and-burn deforestation to clear land for agriculture.

Reducing emissions from deforestation and forest degradation (REDD) in developing countries has emerged as a new potential to complement ongoing climate policies. The idea consists in providing financial compensations for the reduction of greenhouse gas (GHG) emissions from deforestation and forest degradation".

Rainforests are widely believed by laymen to contribute a significant amount of the world's oxygen, although it is now accepted by scientists that rainforests contribute little net oxygen to the atmosphere and deforestation has only a minor effect on atmospheric oxygen levels. However, the incineration and burning of forest plants to clear land releases large amounts of CO_2, which contributes to global warming. Scientists also state that tropical deforestation releases 1.5 billion tons of carbon each year into the atmosphere.

Hydrological

The water cycle is also affected by deforestation. Trees extract groundwater through their roots and release it into the atmosphere. When part of a forest is removed, the trees no longer transpire this water, resulting in a much drier climate. Deforestation reduces the content of water in the soil and groundwater as well as atmospheric moisture. The dry soil leads to lower water intake for the trees to extract. Deforestation reduces soil cohesion, so that erosion, flooding and landslides ensue.

Shrinking forest cover lessens the landscape's capacity to intercept, retain and transpire precipitation. Instead of trapping precipitation, which then percolates to groundwater systems, deforested areas become sources of surface water runoff, which moves much faster than subsurface flows. That quicker transport of surface water can translate into flash flooding and more localized floods than would occur with the forest cover. Deforestation also contributes to decreased evapotranspiration, which lessens atmospheric moisture which in some cases affects precipitation levels downwind from the deforested area, as water is not recycled to downwind forests, but is lost in runoff and returns directly to the oceans. According to one study, in deforested north and northwest China, the average annual precipitation decreased by one third between the 1950s and the 1980s.

Trees, and plants in general, affect the water cycle significantly:

- their canopies intercept a proportion of precipitation, which is then evaporated back to the atmosphere (canopy interception);
- their litter, stems and trunks slow down surface runoff;
- their roots create macropores – large conduits – in the soil that increase infiltration of water;
- they contribute to terrestrial evaporation and reduce soil moisture via transpiration;
- their litter and other organic residue change soil properties that affect the capacity of soil to store water.
- their leaves control the humidity of the atmosphere by transpiring. 99% of the water absorbed by the roots moves up to the leaves and is transpired.

As a result, the presence or absence of trees can change the quantity of water on the surface, in the soil or groundwater, or in the atmosphere. This in turn changes erosion rates and the availability of water for either ecosystem functions or human services.

The forest may have little impact on flooding in the case of large rainfall events, which overwhelm the storage capacity of forest soil if the soils are at or close to saturation.

Tropical rainforests produce about 30% of our planet's fresh water.

Soil

Deforestation for the use of clay in the Brazilian city of Rio de Janeiro. The hill depicted is Morro da Covanca, in Jacarepaguá

Undisturbed forests have a very low rate of soil loss (erosion), approximately 2 metric tons per square kilometer (6 short tons per square mile). Deforestation generally increases rates of soil loss, by increasing the amount of runoff and reducing the protection of the soil from tree litter. This can be an advantage in excessively leached tropical rain forest soils. Forestry operations themselves also increase erosion through the development of (forest) roads and the use of mechanized equipment.

China's Loess Plateau was cleared of forest millennia ago. Since then it has been eroding, creating dramatic incised valleys, and providing the sediment that gives the Yellow River its yellow color and that causes the flooding of the river in the lower reaches (hence the river's nickname 'China's sorrow').

Removal of trees does not always increase erosion rates. In certain regions of southwest US, shrubs and trees have been encroaching on grassland. The trees themselves enhance the loss of grass between tree canopies. The bare intercanopy areas become highly erodible. The US Forest Service, in Bandelier National Monument for example, is studying how to restore the former ecosystem, and reduce erosion, by removing the trees.

Tree roots bind soil together, and if the soil is sufficiently shallow they act to keep the soil in place by also binding with underlying bedrock. Tree removal on steep slopes with shallow soil thus increases the risk of landslides, which can threaten people living nearby.

Biodiversity

Deforestation on a human scale results in decline in biodiversity, and on a natural global scale is known to cause the extinction of many species. The removal or destruction of areas of forest cover has resulted in a degraded environment with reduced biodiversity. Forests support biodiversity, providing habitat for wildlife; moreover, forests foster medicinal conservation. With forest biotopes being irreplaceable source of new drugs (such as taxol), deforestation can destroy genetic variations (such as crop resistance) irretrievably.

Illegal logging in Madagascar. In 2009, the vast majority of the illegally obtained rosewood was exported to China.

Since the tropical rainforests are the most diverse ecosystems on Earth and about 80% of the world's known biodiversity could be found in tropical rainforests, removal or destruction of significant areas of forest cover has resulted in a degraded environment with reduced biodiversity. A study in Rondônia, Brazil, has shown that deforestation also removes the microbial community which is involved in the recycling of nutrients, the production of clean water and the removal of pollutants.

It has been estimated that we are losing 137 plant, animal and insect species every single day due to rainforest deforestation, which equates to 50,000 species a year. Others state that tropical rainforest deforestation is contributing to the ongoing Holocene mass extinction. The known extinction rates from deforestation rates are very low, approximately 1 species per year from mammals and birds which extrapolates to approximately 23,000 species per year for all species. Predictions have been made that more than 40% of the animal and plant species in Southeast Asia could be wiped out in the 21st century. Such predictions were called into question by 1995 data that show that within regions of Southeast Asia much of the original forest has been converted to monospecific plantations, but that potentially endangered species are few and tree flora remains widespread and stable.

Scientific understanding of the process of extinction is insufficient to accurately make predictions about the impact of deforestation on biodiversity. Most predictions of forestry related biodiversity loss are based on species-area models, with an underlying assumption that as the forest declines species diversity will decline similarly. However, many such models have been proven to be wrong and loss of habitat does not necessarily lead to large scale loss of species. Species-area models are known to overpredict the number of species known to be threatened in areas where actual deforestation is ongoing, and greatly overpredict the number of threatened species that are widespread.

A recent study of the Brazilian Amazon predicts that despite a lack of extinctions thus far, up to 90 percent of predicted extinctions will finally occur in the next 40 years.

Economic Impact

Damage to forests and other aspects of nature could halve living standards for the world's poor and reduce global GDP by about 7% by 2050, a report concluded at the Convention on Biological Diversity (CBD) meeting in Bonn. Historically, utilization of forest products, including timber and fuel wood, has played a key role in human societies, comparable to the roles of water and cultivable land. Today, developed countries continue to utilize timber for building houses, and wood pulp for paper. In developing countries almost three billion people rely on wood for heating and cooking.

The forest products industry is a large part of the economy in both developed and developing countries. Short-term economic gains made by conversion of forest to agriculture, or over-exploitation of wood products, typically leads to loss of long-term income and long-term biological productivity. West Africa, Madagascar, Southeast Asia and many other regions have experienced lower revenue because of declining timber harvests. Illegal logging causes billions of dollars of losses to national economies annually.

The new procedures to get amounts of wood are causing more harm to the economy and overpower the amount of money spent by people employed in logging. According to a study, "in most areas studied, the various ventures that prompted deforestation rarely generated more than US$5 for every ton of carbon they released and frequently returned far less than US$1". The price on the European market for an offset tied to a one-ton reduction in carbon is 23 euro (about US$35).

Rapidly growing economies also have an effect on deforestation. Most pressure will come from the world's developing countries, which have the fastest-growing populations and most rapid economic (industrial) growth. In 1995, economic growth in developing countries reached nearly 6%, compared with the 2% growth rate for developed countries." As our human population grows, new homes, communities, and expansions of cities will occur. Connecting all of the new expansions will be roads, a very important part in our daily life. Rural roads promote economic development but also facilitate deforestation. About 90% of the deforestation has occurred within 100 km of roads in most parts of the Amazon.

The European Union is one of the largest importer of products made from illegal deforestation.

Forest Transition Theory

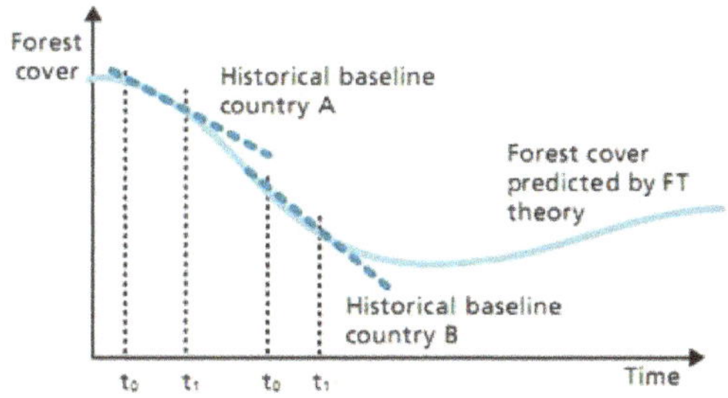

Source: Angelsen 2008.

The forest transition and historical baselines.

The forest area change may follow a pattern suggested by the forest transition (FT) theory, whereby at early stages in its development a country is characterized by high forest cover and low deforestation rates (HFLD countries).

Then deforestation rates accelerate (HFHD, high forest cover – high deforestation rate), and forest cover is reduced (LFHD, low forest cover – high deforestation rate), before the deforestation rate slows (LFLD, low forest cover – low deforestation rate), after which forest cover stabilizes and eventually starts recovering. FT is not a "law of nature," and the pattern is influenced by national context (for example, human population density, stage of development, structure of the economy), global economic forces, and government policies. A country may reach very low levels of forest cover before it stabilizes, or it might through good policies be able to "bridge" the forest transition.

FT depicts a broad trend, and an extrapolation of historical rates therefore tends to underestimate future BAU deforestation for counties at the early stages in the transition (HFLD), while it tends to overestimate BAU deforestation for countries at the later stages (LFHD and LFLD).

Countries with high forest cover can be expected to be at early stages of the FT. GDP per capita captures the stage in a country's economic development, which is linked to the pattern of natural resource use, including forests. The choice of forest cover and GDP per capita also fits well with the two key scenarios in the FT:

(i) a forest scarcity path, where forest scarcity triggers forces (for example, higher prices of forest products) that lead to forest cover stabilization; and

(ii) an economic development path, where new and better off-farm employment opportunities associated with economic growth (= increasing GDP per capita) reduce profitability of frontier agriculture and slows deforestation.

Historical causes

Prehistory

The Carboniferous Rainforest Collapse was an event that occurred 300 million years ago. Climate change devastated tropical rainforests causing the extinction of many plant and animal species. The change was abrupt, specifically, at this time climate became cooler and drier, conditions that are not favourable to the growth of rainforests and much of the biodiversity within them. Rainforests were fragmented forming shrinking 'islands' further and further apart. This sudden collapse affected several large groups, effects on amphibians were particularly devastating, while reptiles fared better, being ecologically adapted to the drier conditions that followed.

An array of Neolithic artifacts, including bracelets, axe heads, chisels, and polishing tools.

Rainforests once covered 14% of the earth's land surface; now they cover a mere 6%

and experts estimate that the last remaining rainforests could be consumed in less than 40 years. Small scale deforestation was practiced by some societies for tens of thousands of years before the beginnings of civilization. The first evidence of deforestation appears in the Mesolithic period. It was probably used to convert closed forests into more open ecosystems favourable to game animals. With the advent of agriculture, larger areas began to be deforested, and fire became the prime tool to clear land for crops. In Europe there is little solid evidence before 7000 BC. Mesolithic foragers used fire to create openings for red deer and wild boar. In Great Britain, shade-tolerant species such as oak and ash are replaced in the pollen record by hazels, brambles, grasses and nettles. Removal of the forests led to decreased transpiration, resulting in the formation of upland peat bogs. Widespread decrease in elm pollen across Europe between 8400–8300 BC and 7200–7000 BC, starting in southern Europe and gradually moving north to Great Britain, may represent land clearing by fire at the onset of Neolithic agriculture.

The Neolithic period saw extensive deforestation for farming land. Stone axes were being made from about 3000 BC not just from flint, but from a wide variety of hard rocks from across Britain and North America as well. They include the noted Langdale axe industry in the English Lake District, quarries developed at Penmaenmawr in North Wales and numerous other locations. Rough-outs were made locally near the quarries, and some were polished locally to give a fine finish. This step not only increased the mechanical strength of the axe, but also made penetration of wood easier. Flint was still used from sources such as Grimes Graves but from many other mines across Europe.

Evidence of deforestation has been found in Minoan Crete; for example the environs of the Palace of Knossos were severely deforested in the Bronze Age.

Pre-industrial History

Throughout most of history, humans were hunter gatherers who hunted within forests. In most areas, such as the Amazon, the tropics, Central America, and the Caribbean, only after shortages of wood and other forest products occur are policies implemented to ensure forest resources are used in a sustainable manner.

In ancient Greece, Tjeered van Andel and co-writers summarized three regional studies of historic erosion and alluviation and found that, wherever adequate evidence exists, a major phase of erosion follows, by about 500-1,000 years the introduction of farming in the various regions of Greece, ranging from the later Neolithic to the Early Bronze Age. The thousand years following the mid-first millennium BC saw serious, intermittent pulses of soil erosion in numerous places. The historic silting of ports along the southern coasts of Asia Minor (*e.g.* Clarus, and the examples of Ephesus, Priene and Miletus, where harbors had to be abandoned because of the silt deposited by the Meander) and in coastal Syria during the last centuries BC.

Easter Island

Easter Island has suffered from heavy soil erosion in recent centuries, aggravated by agriculture and deforestation. Jared Diamond gives an extensive look into the collapse of the ancient Easter Islanders in his book *Collapse*. The disappearance of the island's trees seems to coincide with a decline of its civilization around the 17th and 18th century. He attributed the collapse to deforestation and over-exploitation of all resources.

The famous silting up of the harbor for Bruges, which moved port commerce to Antwerp, also followed a period of increased settlement growth (and apparently of deforestation) in the upper river basins. In early medieval Riez in upper Provence, alluvial silt from two small rivers raised the riverbeds and widened the floodplain, which slowly buried the Roman settlement in alluvium and gradually moved new construction to higher ground; concurrently the headwater valleys above Riez were being opened to pasturage.

A typical progress trap was that cities were often built in a forested area, which would provide wood for some industry (for example, construction, shipbuilding, pottery). When deforestation occurs without proper replanting, however; local wood supplies become difficult to obtain near enough to remain competitive, leading to the city's abandonment, as happened repeatedly in Ancient Asia Minor. Because of fuel needs, mining and metallurgy often led to deforestation and city abandonment.

With most of the population remaining active in (or indirectly dependent on) the agricultural sector, the main pressure in most areas remained land clearing for crop and cattle farming. Enough wild green was usually left standing (and partially used, for example, to collect firewood, timber and fruits, or to graze pigs) for wildlife to remain viable. The elite's (nobility and higher clergy) protection of their own hunting privileges and game often protected significant woodlands.

Major parts in the spread (and thus more durable growth) of the population were played by monastical 'pioneering' (especially by the Benedictine and Commercial orders) and some feudal lords' recruiting farmers to settle (and become tax payers) by offering relatively good legal and fiscal conditions. Even when speculators sought to encourage towns, settlers needed an agricultural belt around or sometimes within defensive walls. When populations were quickly decreased by causes such as the Black Death or devastating warfare (for example, Genghis Khan's Mongol hordes in eastern and central Europe, Thirty Years' War in Germany), this could lead to settlements being abandoned. The land was reclaimed by nature, but the secondary forests usually lacked the original biodiversity.

Deforestation of Brazil's Atlantic Forest c.1820-1825

From 1100 to 1500 AD, significant deforestation took place in Western Europe as a result of the expanding human population. The large-scale building of wooden sailing ships by European (coastal) naval owners since the 15th century for exploration, colonisation, slave trade–and other trade on the high seas consumed many forest resources. Piracy also contributed to the over harvesting of forests, as in Spain. This led to a weakening of the domestic economy after Columbus' discovery of America, as the economy became dependent on colonial activities (plundering, mining, cattle, plantations, trade, etc.)

In *Changes in the Land* (1983), William Cronon analyzed and documented 17th-century English colonists' reports of increased seasonal flooding in New England during the period when new settlers initially cleared the forests for agriculture. They believed flooding was linked to widespread forest clearing upstream.

The massive use of charcoal on an industrial scale in Early Modern Europe was a new type of consumption of western forests; even in Stuart England, the relatively primitive production of charcoal has already reached an impressive level. Stuart England was so widely deforested that it depended on the Baltic trade for ship timbers, and looked to the untapped forests of New England to supply the need. Each of Nelson's Royal Navy war ships at Trafalgar (1805) required 6,000 mature oaks for its construction. In France, Colbert planted oak forests to supply the French navy in the future. When the oak plantations matured in the mid-19th century, the masts were no longer required because shipping had changed.

Norman F. Cantor's summary of the effects of late medieval deforestation applies equally well to Early Modern Europe:

Europeans had lived in the midst of vast forests throughout the earlier medieval centuries. After 1250 they became so skilled at deforestation that by 1500 they were running short of wood for heating and cooking. They were faced with a nutritional decline because of the elimination of the generous supply of wild game that had inhabited the now-disappearing forests, which throughout medieval times had provided the staple of their carnivorous high-protein diet. By 1500 Europe was on the edge of a fuel and nutritional disaster [from] which it was saved in the sixteenth century only by the burning of soft coal and the cultivation of potatoes and maize.

Industrial Era

In the 19th century, introduction of steamboats in the United States was the cause of deforestation of banks of major rivers, such as the Mississippi River, with increased and more severe flooding one of the environmental results. The steamboat crews cut wood every day from the riverbanks to fuel the steam engines. Between St. Louis and the confluence with the Ohio River to the south, the Mississippi became more wide and shallow, and changed its channel laterally. Attempts to improve navigation by the use of snag pullers often resulted in crews' clearing large trees 100 to 200 feet (61 m) back from the banks. Several French colonial towns of the Illinois Country, such as Kaskaskia, Cahokia and St. Philippe, Illinois were flooded and abandoned in the late 19th century, with a loss to the cultural record of their archeology.

The wholescale clearance of woodland to create agricultural land can be seen in many parts of the world, such as the Central forest-grasslands transition and other areas of the Great Plains of the United States. Specific parallels are seen in the 20th-century deforestation occurring in many developing nations.

Rates of Deforestation

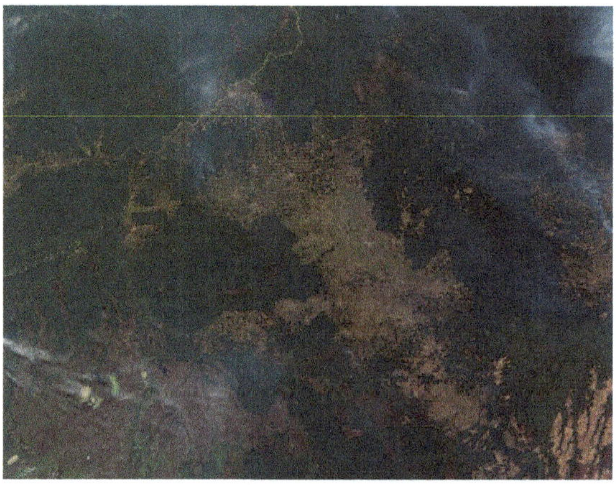

Slash-and-burn farming in the state of Rondônia, western Brazil

Global deforestation sharply accelerated around 1852. It has been estimated that about half of the Earth's mature tropical forests—between 7.5 million and 8 million km² (2.9

million to 3 million sq mi) of the original 15 million to 16 million km² (5.8 million to 6.2 million sq mi) that until 1947 covered the planet—have now been destroyed. Some scientists have predicted that unless significant measures (such as seeking out and protecting old growth forests that have not been disturbed) are taken on a worldwide basis, by 2030 there will only be 10% remaining, with another 10% in a degraded condition. 80% will have been lost, and with them hundreds of thousands of irreplaceable species. Some cartographers have attempted to illustrate the sheer scale of deforestation by country using a cartogram.

Estimates vary widely as to the extent of tropical deforestation. Scientists estimate that one fifth of the world's tropical rainforest was destroyed between 1960 and 1990. They claim that that rainforests 60 years ago covered 14% of the world's land surface, now only cover 5–7%, and that all tropical forests will be gone by the middle of the 21st century.

A 2002 analysis of satellite imagery suggested that the rate of deforestation in the humid tropics (approximately 5.8 million hectares per year) was roughly 23% lower than the most commonly quoted rates. Conversely, a newer analysis of satellite images reveals that deforestation of the Amazon rainforest is twice as fast as scientists previously estimated.

Some have argued that deforestation trends may follow a Kuznets curve, which if true would nonetheless fail to eliminate the risk of irreversible loss of non-economic forest values (for example, the extinction of species).

Satellite image of Haiti's border with the Dominican Republic (right) shows the amount of deforestation on the Haitian side

A 2005 report by the United Nations Food and Agriculture Organization (FAO) estimates that although the Earth's total forest area continues to decrease at about 13 million hectares per year, the global rate of deforestation has recently been slowing. Still others claim that rainforests are being destroyed at an ever-quickening pace. The London-based Rainforest Foundation notes that "the UN figure is based on a definition of forest as being an area with as little as 10% actual tree cover, which would therefore include areas that are actually savannah-like ecosystems and badly damaged forests." Other critics of the FAO data point out that they do not distinguish between forest types, and that they are based largely on reporting from forestry departments of individual countries, which do not take into account unofficial activities like illegal logging.

Despite these uncertainties, there is agreement that destruction of rainforests remains a significant environmental problem. Up to 90% of West Africa's coastal rainforests have disappeared since 1900. In South Asia, about 88% of the rainforests have been lost. Much of what remains of the world's rainforests is in the Amazon basin, where the Amazon Rainforest covers approximately 4 million square kilometres. The regions with the highest tropical deforestation rate between 2000 and 2005 were Central America—which lost 1.3% of its forests each year—and tropical Asia. In Central America, two-thirds of lowland tropical forests have been turned into pasture since 1950 and 40% of all the rainforests have been lost in the last 40 years. Brazil has lost 90–95% of its Mata Atlântica forest. Paraguay was losing its natural semi humid forests in the country's western regions at a rate of 15.000 hectares at a randomly studied 2-month period in 2010, Paraguay's parliament refused in 2009 to pass a law that would have stopped cutting of natural forests altogether.

Deforestation around Pakke Tiger Reserve, India

Madagascar has lost 90% of its eastern rainforests. As of 2007, less than 1% of Haiti's forests remained. Mexico, India, the Philippines, Indonesia, Thailand, Burma, Malaysia, Bangladesh, China, Sri Lanka, Laos, Nigeria, the Democratic Republic of the Congo, Liberia, Guinea, Ghana and the Ivory Coast, have lost large areas of their rainforest. Several countries, notably Brazil, have declared their deforestation a national emergency. The World Wildlife Fund's ecoregion project catalogues habitat types throughout the world, including habitat loss such as deforestation, showing for example that even in the rich forests of parts of Canada such as the Mid-Continental Canadian forests of the prairie provinces half of the forest cover has been lost or altered.

Regions

Rates of deforestation vary around the world.

In 2011 Conservation International listed the top 10 most endangered forests, characterized by having all lost 90% or more of their original habitat, and each harboring at least 1500 endemic plant species (species found nowhere else in the world).

Top 10 Most Endangered Forests 2011				
Endangered forest	**Region**	**Remaining habitat**	**Predominate vegetation type**	**Notes**
Indo-Burma	Asia-Pacific	5%	Tropical and subtropical moist broadleaf forests	Rivers, floodplain wetlands, mangrove forests. Burma, Thailand, Laos, Vietnam, Cambodia, India.
New Caledonia	Asia-Pacific	5%	Tropical and subtropical moist broadleaf forests	See note for region covered.
Sundaland	Asia-Pacific	7%	Tropical and subtropical moist broadleaf forests	Western half of the Indo-Malayan archipelago including southern Borneo and Sumatra.
Philippines	Asia-Pacific	7%	Tropical and subtropical moist broadleaf forests	Forests over the entire country including 7,100 islands.
Atlantic Forest	South America	8%	Tropical and subtropical moist broadleaf forests	Forests along Brazil's Atlantic coast, extends to parts of Paraguay, Argentina and Uruguay.
Mountains of Southwest China	Asia-Pacific	8%	Temperate coniferous forest	See note for region covered.
California Floristic Province	North America	10%	Tropical and subtropical dry broadleaf forests	See note for region covered.
Coastal Forests of Eastern Africa	Africa	10%	Tropical and subtropical moist broadleaf forests	Mozambique, Tanzania, Kenya, Somalia.
Madagascar & Indian Ocean Islands	Africa	10%	Tropical and subtropical moist broadleaf forests	Madagascar, Mauritius, Reunion, Seychelles, Comoros.
Eastern Afromontane	Africa	11%	Tropical and subtropical moist broadleaf forests Montane grasslands and shrublands	Forests scattered along the eastern edge of Africa, from Saudi Arabia in the north to Zimbabwe in the south.

Control

Reducing Emissions

Main international organizations including the United Nations and the World Bank, have begun to develop programs aimed at curbing deforestation. The blanket term Reducing Emissions from Deforestation and Forest Degradation (REDD) describes these sorts of programs, which use direct monetary or other incentives to encourage developing countries to limit and/or roll back deforestation. Funding has been an issue, but at the UN Framework Convention on Climate Change (UNFCCC) Conference of the Parties-15 (COP-15) in Copenhagen in December 2009, an accord was reached with a collective commitment by developed countries for new and additional resources, including forestry and investments through international institutions, that will approach USD 30 billion for the period 2010–2012. Significant work is underway on tools for use in monitoring developing country adherence to their agreed REDD targets. These tools, which rely on remote forest monitoring using satellite imagery and other data sources, include the Center for Global Development's FORMA (Forest Monitoring for Action) initiative and the Group on Earth Observations' Forest Carbon Tracking Portal. Methodological guidance for forest monitoring was also emphasized at COP-15. The environmental organization Avoided Deforestation Partners leads the campaign for development of REDD through funding from the U.S. government. In 2014, the Food and Agriculture Organization of the United Nations and partners launched Open Foris - a set of open-source software tools that assist countries in gathering, producing and disseminating information on the state of forest resources. The tools support the inventory lifecycle, from needs assessment, design, planning, field data collection and management, estimation analysis, and dissemination. Remote sensing image processing tools are included, as well as tools for international reporting for Reducing emissions from deforestation and forest degradation (REDD) and MRV and FAO's Global Forest Resource Assessments.

In evaluating implications of overall emissions reductions, countries of greatest concern are those categorized as High Forest Cover with High Rates of Deforestation (HFHD) and Low Forest Cover with High Rates of Deforestation (LFHD). Afghanistan, Benin, Botswana, Burma, Burundi, Cameroon, Chad, Ecuador, El Salvador, Ethiopia, Ghana, Guatemala, Guinea, Haiti, Honduras, Indonesia, Liberia, Malawi, Mali, Mauritania, Mongolia, Namibia, Nepal, Nicaragua, Niger, Nigeria, Pakistan, Paraguay, Philippines, Senegal, Sierra Leone, Sri Lanka, Sudan, Togo, Uganda, United Republic of Tanzania, Zimbabwe are listed as having Low Forest Cover with High Rates of Deforestation (LFHD). Brazil, Cambodia, Democratic Peoples Republic of Korea, Equatorial Guinea, Malaysia, Solomon Islands, Timor-Leste, Venezuela, Zambia are listed as High Forest Cover with High Rates of Deforestation (HFHD).

Payments for Conserving Forests

In Bolivia, deforestation in upper river basins has caused environmental problems, including soil erosion and declining water quality. An innovative project to try and remedy this situation involves landholders in upstream areas being paid by downstream water users to conserve forests. The landholders receive US$20 to conserve the trees, avoid polluting livestock practices, and enhance the biodiversity and forest carbon on their land. They also receive US$30, which purchases a beehive, to compensate for conservation for two hectares of water-sustaining forest for five years. Honey revenue per hectare of forest is US$5 per year, so within five years, the landholder has sold US$50 of honey. The project is being conducted by Fundación Natura Bolivia and Rare Conservation, with support from the Climate & Development Knowledge Network.

Farming

New methods are being developed to farm more intensively, such as high-yield hybrid crops, greenhouse, autonomous building gardens, and hydroponics. These methods are often dependent on chemical inputs to maintain necessary yields. In cyclic agriculture, cattle are grazed on farm land that is resting and rejuvenating. Cyclic agriculture actually increases the fertility of the soil. Intensive farming can also decrease soil nutrients by consuming at an accelerated rate the trace minerals needed for crop growth. The most promising approach, however, is the concept of food forests in permaculture, which consists of agroforestal systems carefully designed to mimic natural forests, with an emphasis on plant and animal species of interest for food, timber and other uses. These systems have low dependence on fossil fuels and agro-chemicals, are highly self-maintaining, highly productive, and with strong positive impact on soil and water quality, and biodiversity.

Monitoring Deforestation

There are multiple methods that are appropriate and reliable for reducing and monitoring deforestation. One method is the "visual interpretation of aerial photos or satellite imagery that is labor-intensive but does not require high-level training in computer image processing or extensive computational resources". Another method includes hot-spot analysis (that is, locations of rapid change) using expert opinion or coarse resolution satellite data to identify locations for detailed digital analysis with high resolution satellite images. Deforestation is typically assessed by quantifying the amount of area deforested, measured at the present time. From an environmental point of view, quantifying the damage and its possible consequences is a more important task, while conservation efforts are more focused on forested land protection and development of land-use alternatives to avoid continued deforestation. Deforestation rate and total area deforested, have been widely used for monitoring deforestation in many regions, including the Brazilian Amazon deforestation monitoring by INPE.

Forest Management

Efforts to stop or slow deforestation have been attempted for many centuries because it has long been known that deforestation can cause environmental damage sufficient in some cases to cause societies to collapse. In Tonga, paramount rulers developed policies designed to prevent conflicts between short-term gains from converting forest to farmland and long-term problems forest loss would cause, while during the 17th and 18th centuries in Tokugawa, Japan, the shoguns developed a highly sophisticated system of long-term planning to stop and even reverse deforestation of the preceding centuries through substituting timber by other products and more efficient use of land that had been farmed for many centuries. In 16th-century Germany, landowners also developed silviculture to deal with the problem of deforestation. However, these policies tend to be limited to environments with *good rainfall*, *no dry season* and *very young soils* (through volcanism or glaciation). This is because on older and less fertile soils trees grow too slowly for silviculture to be economic, whilst in areas with a strong dry season there is always a risk of forest fires destroying a tree crop before it matures.

In the areas where "slash-and-burn" is practiced, switching to "slash-and-char" would prevent the rapid deforestation and subsequent degradation of soils. The biochar thus created, given back to the soil, is not only a durable carbon sequestration method, but it also is an extremely beneficial amendment to the soil. Mixed with biomass it brings the creation of terra preta, one of the richest soils on the planet and the only one known to regenerate itself.

Sustainable Practices

Bamboo is advocated as a more sustainable alternative for cutting down wood for fuel.

Certification, as provided by global certification systems such as Programme for the Endorsement of Forest Certification and Forest Stewardship Council, contributes to tackling deforestation by creating market demand for timber from sustainably managed forests. According to the United Nations Food and Agriculture Organization (FAO), "A major condition for the adoption of sustainable forest management is a demand for products that are produced sustainably and consumer willingness to pay for the higher costs entailed. Certification represents a shift from regulatory approaches to market incentives to promote sustainable forest management. By promoting the positive attributes of forest products from sustainably managed forests, certification focuses on the demand side of environmental conservation." Rainforest Rescue argues that the standards of organizations like FSC are too closely connected to timber industry interests and therefore do not guarantee environmentally and socially responsible forest management. In reality, monitoring systems are inadequate and various cases of fraud have been documented worldwide.

Some nations have taken steps to help increase the amount of trees on Earth. In 1981, China created National Tree Planting Day Forest and forest coverage had now reached 16.55% of China's land mass, as against only 12% two decades ago

Using fuel from bamboo rather than wood results in cleaner burning, and since bamboo matures much faster than wood, deforestation is reduced as supply can be replenished faster.

Reforestation

In many parts of the world, especially in East Asian countries, reforestation and afforestation are increasing the area of forested lands. The amount of woodland has increased in 22 of the world's 50 most forested nations. Asia as a whole gained 1 million hectares of forest between 2000 and 2005. Tropical forest in El Salvador expanded more than 20% between 1992 and 2001. Based on these trends, one study projects that global forest will increase by 10%—an area the size of India—by 2050.

In the People's Republic of China, where large scale destruction of forests has occurred, the government has in the past required that every able-bodied citizen between the ages of 11 and 60 plant three to five trees per year or do the equivalent amount of work in other forest services. The government claims that at least 1 billion trees have been planted in China every year since 1982. This is no longer required today, but March 12 of every year in China is the Planting Holiday. Also, it has introduced the Green Wall of China project, which aims to halt the expansion of the Gobi desert through the planting of trees. However, due to the large percentage of trees dying off after planting (up to 75%), the project is not very successful. There has been a 47-million-hectare increase in forest area in China since the 1970s. The total number of trees amounted to be about 35 billion and 4.55% of China's land mass increased in forest coverage. The forest coverage was 12% two decades ago and now is 16.55%.

An ambitious proposal for China is the Aerially Delivered Re-forestation and Erosion Control System and the proposed Sahara Forest Project coupled with the Seawater Greenhouse.

In Western countries, increasing consumer demand for wood products that have been produced and harvested in a sustainable manner is causing forest landowners and forest industries to become increasingly accountable for their forest management and timber harvesting practices.

The Arbor Day Foundation's Rain Forest Rescue program is a charity that helps to prevent deforestation. The charity uses donated money to buy up and preserve rainforest land before the lumber companies can buy it. The Arbor Day Foundation then protects the land from deforestation. This also locks in the way of life of the primitive tribes living on the forest land. Organizations such as Community Forestry International, Cool Earth, The Nature Conservancy, World Wide Fund for Nature, Conservation International, African Conservation Foundation and Greenpeace also focus on preserving forest habitats. Greenpeace in particular has also mapped out the forests that are still intact and published this information on the internet. World Resources Institute in turn has made a simpler thematic map showing the amount of forests present just before the age of man (8000 years ago) and the current (reduced) levels of forest. These maps mark the amount of afforestation required to repair the damage caused by people.

Forest Plantations

To meet the world's demand for wood, it has been suggested by forestry writers Botkins and Sedjo that high-yielding forest plantations are suitable. It has been calculated that plantations yielding 10 cubic meters per hectare annually could supply all the timber required for international trade on 5% of the world's existing forestland. By contrast, natural forests produce about 1–2 cubic meters per hectare; therefore, 5–10 times more forestland would be required to meet demand. Forester Chad Oliver has suggested a forest mosaic with high-yield forest lands interspersed with conservation land.

In the country of Senegal, on the western coast of Africa, a movement headed by youths has helped to plant over 6 million mangrove trees. The trees will protect local villages from storm damages and will provide a habitat for local wildlife. The project started in 2008, and already the Senegalese government has been asked to establish rules and regulations that would protect the new mangrove forests.

Military Context

American Sherman tanks knocked out by Japanese artillery on Okinawa.

While the preponderance of deforestation is due to demands for agricultural and urban use for the human population, there are some examples of military causes. One example of deliberate deforestation is that which took place in the U.S. zone of occupation in Germany after World War II. Before the onset of the Cold War, defeated Germany was still considered a potential future threat rather than potential future ally. To address this threat, attempts were made to lower German industrial potential, of which forests were deemed an element. Sources in the U.S. government admitted that the purpose of this was that the "ultimate destruction of the war potential of German forests." As a consequence of the practice of clear-felling, deforestation resulted which could "be replaced only by long forestry development over perhaps a century."

Deforestation can also be one consequence of war. For example, in the 1945 Battle of Okinawa, bombardment and other combat operations reduced the lush tropical landscape into "a vast field of mud, lead, decay and maggots". Deforestation can also be an intentional tactic of military forces. Defoliants (like Agent Orange or others) was used by the British in the Malayan Emergency, and by the United States in the Korean War and Vietnam War.

Public Health Context

Deforestation eliminates a great number of species of plants and animals which also often results in an increase in disease. Loss of native species allows new species to come to dominance. Often the destruction of predatory species can result in an increase in rodent populations. These are known to carry plagues. Additionally, erosion can produce pools of stagnant water that are perfect breeding grounds for mosquitos, well known vectors of malaria, yellow fever, nipah virus, and more. Deforestation can also create a path for non-native species to flourish such as certain types of snails, which have been correlated with an increase in schistosomiasis cases.

Deforestation is occurring all over the world and has been coupled with an increase in the occurrence of disease outbreaks. In Malaysia, thousands of acres of forest have been cleared for pig farms. This has resulted in an increase in the zoonosis the Nipah virus. In Kenya, deforestation has led to an increase in malaria cases which is now the leading cause of morbidity and mortality the country.

Another pathway through which deforestation affects disease is the relocation and dispersion of disease-carrying hosts. This disease emergence pathway can be called "range expansion," whereby the host's range (and thereby the range of pathogens) expands to new geographic areas. Through deforestation, hosts and reservoir species are forced into neighboring habitats. Accompanying the reservoir species are pathogens that have the ability to find new hosts in previously unexposed regions. As these pathogens and species come into closer contact with humans, they are infected both directly and indirectly.

A catastrophic example of range expansion is the 1998 outbreak of Nipah Virus in Malaysia. For a number of years, deforestation, drought, and subsequent fires led to a dramatic geographic shift and density of fruit bats, a reservoir for Nipah virus. Deforestation reduced the available fruiting trees in the bats' habitat, and they encroached on surrounding orchards which also happened to be the location of a large number of pigsties. The bats, through proximity spread the Nipah to pigs. While the virus infected the pigs, mortality was much lower than among humans, making the pigs a virulent host leading to the transmission of the virus to humans. This resulted in 265 reported cases of encephalitis, of which 105 resulted in death. This example provides an important lesson for the impact deforestation can have on human health.

Another example of range expansion due to deforestation and other anthropogenic habitat impacts includes the Capybara rodent in Paraguay. This rodent is the host of a number of zoonotic diseases and, while there has not yet been a human-borne outbreak due to the movement of this rodent into new regions, it offers an example of how habitat destruction through deforestation and subsequent movements of species is occurring regularly.

A now well-developed theory is that the spread of HIV it is at least partially due deforestation. Rising populations created a food demand and with deforestation opening up new areas of the forest the hunters harvested a great deal of primate bushmeat, which is believed to be the origin of HIV.

Forest Dieback

Jizera Mountains in Central Europe

Forest dieback (also "Waldsterben", a German loan word) is a condition in trees or woody plants in which peripheral parts are killed, either by pathogens, parasites or due to conditions like acid rain and drought. Two of the nine tipping points for major climate changes, forecast for the next century, are directly related to forest diebacks. *Phomopsis azadirachtae* is a fungus of the Phomopsis genus that has been identified as the fungus responsible for dieback of *Azadirachta indica* (Neem) in India.

Definition

Forest dieback refers to the phenomena of a stand of trees losing health and dying without an obvious cause. This condition is also known as forest decline, forest damage, canopy level dieback, and stand level dieback (also Waldsterben and Waldschäden, German loan words). This usually affects individual species of trees, but can also affect multiple species. Trees and woody plants are killed by a complex group of factors that include parasites, such as fungus and beetles, pollution, such as acid rain and organic compounds, and drought. Diseases and pests can kill large groups of trees very easily, but the premature and progressive loss of clusters of trees without an obvious or specific cause is known as forest dieback. Dieback is an episodic event. Dieback takes on many locations and shapes. It can be along the perimeter, at specific elevations, or dispersed throughout the forest ecosystem.

Forest dieback has many symptoms: falling off of leaves and needles, discoloration of leaves and needles, thinning of the crowns of trees, dead stands of trees of a certain age, and changes in the roots of the trees. Forest dieback has many dynamic forms. A stand of trees can exhibit mild symptoms, extreme systems, or death at the same time. Forest decline can be viewed as the result of continued, widespread, and severe dieback of multiple species in the forest. Current forest decline is described by rapid development on individual trees, occurrence in different forest types, a long duration (over 10 years), and occurrence throughout the natural range of affected species.

History

Much research was done in the 1980s when a severe dieback happened in Germany and the Northeast United States. Previous diebacks were regionally limited, but starting at the end of the 1970s a decline took over the forests of Central Europe and parts of North America. The forest damage in Germany was different because the decline was severe, up to 50% of the trees, prolonged, over 5 years, and widespread across tree species. Affected trees went from 8% in 1982 to 50% in 1984 and stayed at 50% through 1987. Many hypotheses have been proposed for this dieback.

Drought and temperature have been causal factors in many diebacks. This condition is common in semi-arid regions and the trees are already stressed by dehydration. There have been reported diebacks in the 50% mortality range across multiple species in Africa, Asia, Europe, North America, South and Central America, and Australia. Notable drought related diebacks have occurred globally from the 1970s through the 2000s. A few drought related diebacks have occurred before this. For instance, a 23% mortality rate was reported between 1945 and 1993 in Senegal Africa. Trees stressed by drought and temperature are more susceptible to parasites.

In North America there have been five notable hardwood diebacks this century. These have occurred with the maturation of the forest and each episode has lasted about eleven years. The most severe temperate forest dieback was on white birch and yellow trees. The others are ash, oak, and maple species. White and yellow birches experienced an episode starting between 1934 and 1957 and ending between 1953 and 1954. This followed a wave pattern that first appeared in southern regions and moved to northern regions. A second wave is evident between 1957 and 1965 in northern Quebec. Sugar maple experienced a wave of dieback in parts of the United States during the 1960s. A second wave occurred primarily in Canada during the 1980s, but also reached the United States. These diebacks were numerically analyzed to exclude natural tree mortality. It is hypothesized that a mature forest is more susceptible to extreme environmental stresses.

Pathogens are responsible for many diebacks. It is difficult to isolate and identify exactly which pathogens are responsible and how they interact with the trees. For instance *Phomopsis azadirachtae* Sateesh, Bhat, and Devaki are fungus of the *Phomopsis* genus that has been identified as the fungus responsible for the dieback in *Azadirachta indica* (Neem) in the regions of India. Neem is a very hardy and drought tolerant tree, has antifungal and antibacterial properties, but it still has many diseases. Different pathogens affect every part of the tree: the twigs, shoots, leaves, roots, and bark. Scientists have isolated both the fungus and the toxin. They have identified the pathogen in the tissue of all affected trees. Proper knowledge of toxin chemistry and its role in pathogenesis requires further investigations. Some experts consider dieback as a group of diseases with incompletely understood origins influenced by factors which predispose trees under stress to invasion.

Top dying disease is a disease of *Heritiera fomes*, the dominant species of mangrove tree growing in the Sundarbans. This disease has become more prevalent since about 1970 and may be linked to an increase in the heavy metal concentration of the sediment deposited in the Ganges Delta.

Hypothesis

The components of a forest ecosystem are complex and identifying specific cause-effect relationships between dieback and the environment is difficult. Etiology is the science of identifying the causes of death. Because there is no one single and clear cause of dieback there are multiple hypothesis for the causes and effects of dieback. The following hypotheses were agreed upon between the scientific exchanges of Germany and the United States in 1988.

Soil-Acidification/Aluminum-Toxicity: As soil becomes more acidic there is a release of aluminum that damages the tree's roots. Observed effects are: reduction of uptake and transport of some cations, reduction in root respiration, damage to fine feeder roots and root morphology, and reduction in elasticity of the cell walls. This is proposed by Professor Bernhard Ulrich in 1979.

Complex High-Elevation Disease: The combination of high ozone levels, acid deposition and nutrient deficiencies at high elevations kill trees. High ozone concentrations damage the leaves and needles of trees. Nutrients are leached from the foliage. The chain of events is magnified over time. This is proposed by a group of professors: Bernhard Prinz, Karl Rehfuess, Heinz Zöttle.

Red-Needle Disease of Spruce: The symptoms are needle drop and crown thinning. Needles turn a rust color and fall off. This is caused by foliar fungi that are secondary parasites attacking weakened trees. This is proposed by Professor Karl Rehfuess.

General Stress: The increased level of atmospheric pollutants hurts the root system and accumulates toxins in new leaves. Pollutants alter the growth, reduce the photoshythetic activity, and reduce the formation of secondary metabolites. It is believed that low concentrations levels are toxic. This is proposed by a group of professors led by Peter Schütt.

Excess Nitrogen Deposition: The increased level of nitrogen and ammonium, a fertilizer, could have the following possible effects: it could inhibit beneficial fungi, delay chemical reactions, disurbe normal balances between shoot growth and root growth, increase soil leaching. There is no experimental proof. This is proposed by Carl Olaf Tamm.

Organic Air Pollutants: This is similar to the general stress hypothesis, but focuses on organic compounds. The three compounds seriously discussed are ethylene, aniline, and dinitrophenol. Even at low levels these organic chemical compounds have caused: abnormal dropping of foliage, twisted foliage, and killing of seedlings. This is proposed by Fritz Führ.

Global Climate Change

Changes in mean annual temperature and drought are major contributing factors to forest dieback. As more carbon is released from dead trees, especially in the Amazon and Boreal forests, more greenhouse gases are released into the atmosphere. Increased levels of greenhouse gases increase the temperature of the atmosphere. The feedback loop is reinforced and the biological adaptations of the species determine its survival. Projections for dieback vary, but the threat of global climate change only stands to increase the rate of dieback.

Scientists do not know the exact tipping points of climate change and can only estimate the timescales. When a tipping point- the critical threshold- is reached, a small change in human activity can have long-term consequences on the environment. Two of the nine tipping points for major climate changes forecast for the next century are directly related to forest diebacks. Scientists are worried that forest dieback in the Amazon rain forest and the Boreal evergreen forest will trigger a tipping point in the next 50 years.

Illegal Logging

Illegal logging is the harvest, transportation, purchase or sale of timber in violation of laws. The harvesting procedure itself may be illegal, including using corrupt means to gain access to forests; extraction without permission or from a protected area; the cutting of protected species; or the extraction of timber in excess of agreed limits (see Box 1).

Illegalities may also occur during transport, such as illegal processing and export; fraudulent declaration to customs; the avoidance of taxes and other charges, and fraudulent certification.

Overview

Box 1. Logging in national parks: the case of Korindo (Indonesia)

In March 2004, Greenpeace carried out actions against a cargo ship transporting timber from the Indonesian company Korindo, which was being imported into France, UK, Belgium and the Netherlands.

Korindo is known to be using illegal timber from the last rainforests of Indonesia. In May 2003, an Indonesian Government investigation confirmed that Korindo was receiving illegal timber from notorious timber barons known to obtain timber from an orang-utan refuge – the Tanjung Puting National Park. Tanjung Puting National Park is a 4,000 square kilometre conservation area of global importance. It is recognised as a world biosphere reserve by the United Nations and forms the largest protected area of swamp forest in South-East Asia.

Illegal logging is a pervasive problem, causing enormous damage to forests, local communities and to the economies of producer countries. Despite the economic importance of trade in timber and forest products, major international timber consumer countries, such as the EU, have no legal means to halt the import of illegally sourced forest products, because the identification of illegally logged or traded timber is technically difficult. Therefore, a legal basis for normative acts against timber imports or other products manufactured out of illegal wood is missing. Scientific methods to pinpoint the geographic origin of timber are currently under development. Possible actions to restrict imports cannot meet with WTO regulations of non-discrimination. They must instead be arranged in bilateral agreements. TRAFFIC, the wildlife trade monitoring network strives to monitor the illegal trade of timber and provide expertise in policy and legal reviews.

Scale

It is estimated that illegal logging in public lands alone causes losses in assets and revenue in excess of 10 billion USD annually. Although exact figures are difficult to state, given the illegal nature of the activity, decant estimates show that most of the logging, more than half, that is done in the world is illegal most especially in open and vulnerable areas – the Amazon Basin, Central Africa, Southeast Asia, the Russian Federation – is illegal.

Available figures and estimations must be dealt with cautiously. Governments may tend to underestimate the situation. High estimates of illegal logging may constitute an embarrassment as these hint at ineffective enforcement of legislation or, even worse, bribery and corruption. On the other hand, environmental NGOs publish alarming figures to raise awareness and to emphasise the need for stricter conservation measures. For companies of the forest sector, publications providing high estimations can be regarded as potentially threatening for their reputation and their market perspective, including the competitiveness of wood in comparison to other materials. However, for many countries, NGOs are the only source of information apart from state institutions that probably clearly underestimate the situation. For example, the Republic of Estonia calculated an amount of 1% illegally harvested timber in 2003, whereas it is estimated to reach a maximum of 50% by the ENGO "Estonian Green Movement". In Latvia, the situation is comparable; anecdotal evidence points towards 25% of logging being illegal.

Consequences

Illegal logging continues in Thailand. This photograph was taken from the roadside in Mae Wang District, Chiang Mai Province, in March 2011

Illegal logging contributes to deforestation and by extension global warming, causes loss of biodiversity and undermines the rule of law. These illegal activities undermine responsible forest management, encourage corruption and tax evasion and reduce the income of the producer countries, further limiting the resources producer countries can invest in sustainable development. Illegal logging has serious economic and social implications for the poor and disadvantaged with millions of dollars worth of timber revenue being lost each year.

Furthermore, the illegal trade of forest resources undermines international security, and is frequently associated with corruption, money laundering, organized crime, human rights abuses and, in some cases, violent conflict. In the forestry sector, cheap imports of illegal timber and forest products, together with the non-compliance of some economic players with basic social and environmental standards, destabilise international markets. This unfair competition affects those European companies, especially the small and medium-sized companies that are behaving responsibly and ready to play by fair rules.

Statistics on Illegal Logging

Box 2. Loss of revenue to governments of producer countries

The scale of illegal logging represents a major loss of revenue to many countries and can lead to widespread associated environmental damage. A senate committee in the Philippines estimated that the country lost as much as US$1.8bn per year during the 1980s. The Indonesian government estimated in 2002 that costs related to illegal logging are US$3bn each year. The World Bank estimates that illegal logging costs timber-producing countries between 10 and 15 billion euros per year. This compares with 10 billion euros disbursed as EC aid in 2002.

- A joint UK-Indonesian study of the timber industry in Indonesia in 1998 suggested that about 40% of throughput was illegal, with a value in excess of $365 million. More recent estimates, comparing legal harvesting against known domestic consumption plus exports, suggest that 88% of logging in the country is illegal in some way. Malaysia is the key transit country for illegal wood products from Indonesia.

- In Brazil, 80% of logging in the Amazon violates government controls. At the core of illegal logging is widespread corruption. Often referred to as 'green gold', mahogany can fetch over US$1,600 m-3. Illegal mahogany opens the door for illegal logging of other species, and for widespread exploitation of the Brazilian Amazon. Recent Greenpeace investigations in the Brazilian state of Pará reveal just how deeply rooted the problem remains. No reliable legal chain of custody exists for mahogany, and the key players in its trade are ruthless.

- The World Bank estimates that 80% of logging operations are illegal in Bolivia and 42% in Colombia, 10 while in Peru, illegal logging equals 80% of all activities.

- Research carried out by WWF International in 2002 shows that in Africa, rates of illegal logging vary from 50% for Cameroon and Equatorial Guinea to 70% in Gabon and 80% in Liberia – where revenues from the timber industry also fuelled the civil war.

- WWF estimates that illegal logging in Russia is at least 20%, reaching up to 50% in its far eastern regions.

- A 2012 joint study by the United Nations Environment Programme and Interpol states that illegal logging accounts for up to 30% of the global logging trade and contributes to more than 50% of tropical deforestation in Central Africa, the Amazon Basin and South East Asia.

- Between 50% and 90% of logging from the key countries in these regions is being carried out by organised criminal entities.

Political Processes

East Asia

Signs of illegal timber poaching on the boundary of the protected area around the Cagua Volcano, Cagayan, Philippines.

The East Asia Forest Law Enforcement and Governance (EA FLEG) Ministerial Conference took place in Bali in September 2001. The Conference brought together nearly 150 participants from 20 countries, representing government, international organizations, NGOs, and the private sector. The event was co-hosted by the World Bank and the Government of Indonesia.

The meeting included detailed technical discussions of forest law enforcement in relation to governance, forest policy and forest management as well as ministerial engagement.

The Conference's primary aims were to share analysis on forest law enforcement; explore priority issues of forest law enforcement, including illegal logging in the East Asia region, among senior officials from forest and related ministries, NGOs and industry representatives; and commit to action at the national and regional level.

European Union

In May 2003, the European Commission presented the EU Forest Law Enforcement, Governance and Trade Action Plan (EU FLEGT). This marked the beginning of a long process by which the EU aims to develop and implement measures to address illegal logging and related trade. The primary means of implementing the Plan is through Voluntary Partnership Agreements with timber producing countries. The EU Timber Regulation (EUTR) was adopted in 2010 and went into effect 3 March 2013.

- It prohibits the placing on the EU market for the first time of illegally harvested timber and products derived from such timber;
- It requires EU traders who place timber products on the EU market for the first time to exercise 'due diligence';
- Once on the market, the timber and timber products may be sold on and/or transformed before they reach the final consumer.
- To facilitate the traceability of timber products economic operators in this part of the supply chain (referred to as traders in the regulation) have an obligation to keep records of their suppliers and customers.

A Greenpeace investigation published in May 2014 demonstrates that EU Timber Regulation is ineffective if fraudulent paperwork is accepted at face value and there is not sufficient enforcement by EU authorities.

Africa

The *Africa Forest Law Enforcement and Governance* (AFLEG) Ministerial Conference was held in Yaoundé, Cameroon in October 2003. The meeting drew together ministers and stakeholders from Africa, Europe and North America to consider how partnerships between producers, consumers, donors, civil society and the private sector could address illegal forest exploitation and associated trade in Africa.

The AFLEG conference, the second regional forest law enforcement and governance meeting after East Asia, resulted in endorsement of a ministerial declaration and action plan as well as a variety of informal implementation initiatives.

In 2014, the EU-FAO Forest Law Enforcement, Governance and Trade Programme of the Food and Agriculture Organization of the United Nations published the study *The Voluntary Partnership Agreement (VPA) process in Central and West Africa: from theory to practice* to document and foster strategic reflection in partner countries already engaged in negotiating a VPA - or those who will be entering into such negotiations - by providing examples of good practices. These good practices were identified and recorded following interviews with the main stakeholders in the eight VPA countries in West and Central Africa, the European Forest Institute's (EFI) EU FLEGT Facility and the European Commission.

Saint Petersburg Declaration

The *Europe and North Asia Forest Law Enforcement and Governance* (ENA FLEG) Ministerial Conference was held in Saint Petersburg, Russia on 22–25 November 2005. In May 2004, the Russian Federation announced its intention to host the ENA FLEG process, supported by the World Bank. A preparatory conference was held in Moscow in June 2005.

The Saint Petersburg conference brought together nearly 300 participants representing 43 governments, the private sector, civil society and international organisations. It agreed to the *Saint Petersburg Declaration on Forest Law Enforcement and Governance in Europe and North Asia*. The Declaration includes an indicative list of actions, intended to serve as a general framework for possible actions to be undertaken by governments as well as civil society.

The conference took place as the United Kingdom prepared to pass the G8 Presidency to Russia. As Valery Roshchupkin, Head of the Federal Forestry Agency of the Russian Federation, confirmed, illegal logging would be of special importance for Russia as the G8 President and for the following G8 Summit, also held in Saint Petersburg.

United States

In response to growing concerns over illegal logging and advice from TRAFFIC and other organisations, on May 22, 2008 the U.S. amended the Lacey Act, when the Food,

Conservation, and Energy Act of 2008 expanded its protection to a broader range of plants and plant products (Section 8204. Prevention of Illegal Logging Practices).

The requirements under the new Amendments are two-fold. First, the Lacey Act now makes it illegal to import into the United States plants that have been harvested contrary to any applicable Federal Law, State Law, Indian Tribal Law, or Foreign Law. If a plant is found to have been harvested in violation of the laws of the country where it was harvested, that plant would be subject to seizure and forfeiture if imported into the U.S. The Lacey Act also makes it unlawful, beginning December 15, 2008, to import certain plants and plant products without a Plant and Plant Product import declaration.

This Plant and Plant Product Declaration must contain (among other things) the Genus, Species, and Country of Harvest of every plant found in commercial shipments of certain products, a list of applicable products (along with other requirements and guidance) can be found on the USDA APHIS website.

Australia

The Timber Development Association (TDA) welcomes on June 6, 2014's release by the Australian Department of Agriculture of a position paper on the Illegal Logging Prohibition Regulation and guidance on how timber and wood products industry can comply on the Australian Government - Department of Agriculture official website. The release of the Government's guidance coincides with the release of industry developed timber due diligence tools and information through the industry website of Timber Due Diligence.

The Australian Illegal Logging Prohibition Regulation applies to importers into Australia of "regulated timber products" such as sawn timber, wood panels, pulp, paper products, and wood furniture. The Regulation starts on 30 November 2014 and requires that before import of these products or processing of raw logs, due diligence is undertaken to minimise the risk that the timber products or raw logs were illegally logged or incorporate illegally logged timber.

Clearcutting

After a century of clearcutting, this forest, near the source of the Lewis and Clark River in Clatsop County, Oregon, is a patchwork. In each patch, most of the trees are the same age.

Clearcutting, clearfelling, or clearcut logging is a forestry/logging practice in which most or all trees in an area are uniformly cut down. Clearcutting, along with shelter-wood and seed tree harvests, is used by foresters to create certain types of forest ecosystems and to promote select species that require an abundance of sunlight or grow in large, even-age stands. Logging companies and forest-worker unions in some countries support the practice for scientific, safety, and economic reasons. Detractors see clearcutting as synonymous with deforestation, destroying natural habitats and contributing to climate change.

Clearcutting is the most common and economically profitable method of logging. However, clearcutting also imposes other externalities in the form of detrimental side effects such as loss of topsoil; the value of these costs is intensely debated by economic, environmental, and other interests. Aside from the purpose of harvesting wood, clearcutting is also used to create land for farming. The "insatiable human demand for wood and arable land" through clearcutting and other activities has led to the loss of over half of the world's rainforests.

While deforestation of both temperate and tropical rainforests through clearcutting has received considerable media attention in recent years, the other large forests of the world, such as the taiga, also known as boreal forests, are also under threat of rapid development. The same reasons for preserving the world's tropical rainforests also apply to the taiga, as do the reasons for destroying them. In Russia, as in North America and Scandinavia, creating protected areas and granting long-term leases to tend and regenerate trees—thus maximizing future harvests—are ways of limiting the harmful effects of clearcutting. Long-term studies of clearcut forests, such as studies of the Pasoh Rainforest in Malaysia, are also important in providing insights into the preservation of forest resources worldwide.

Types

Many variations of clearcutting exist; the most common professional practices are:

- *Standard (uniform) clearcut* – removal of every stem (whether commercially viable or not), so no canopy remains.
- *Patch clearcut* – removal of all the stems in a limited, predetermined area (patch).
- *Strip clearcut* – removal of all the stems in a row (strip), usually placed perpendicular to the prevailing winds in order to minimize the possibility of windthrow.
- *Clearcutting-with-reserves* – removal of the majority of standing stems save a few reserved for other purposes (for example as snags for wildlife habitat), (often confused with the seed tree method).

- *Slash-and-burn* – the permanent conversion of tropical and subtropicals forests for agricultural purposes. This is most prevalent in tropical and sub-tropical forests in overpopulated regions in developing and least developed countries. Slash-and-burn entails the removal of all stems in a particular area. This can be a form of deforestation, when the land is converted to other uses. However, some indigenous forest peoples, for example the 19th century Forest Finns rotate over the land and it does return to forest and this would be sustainable. Slash and burn techniques are typically used by civilians in search of land for living and agricultural purposes. The forest is first clear cut, and the remaining material is burned. One of the driving forces behind this process is a result of overpopulation and subsequent sprawl. These methods also occur as a result of commercial farming. The lumber is sold for profit, and the land, cleared of all remaining brush and suitable for agricultural development, is sold to farmers.
- *Selection cutting* – which can be done for timber harvesting or for ecological reasons. when so done it is often called ecoforestry.

Clearcutting contrasts with selective cutting, such as high grading, in which only commercially valuable trees are harvested, leaving all others. This practice can reduce the genetic viability of the forest over time, resulting in poorer or less vigorous offspring in the stand. Clearcutting also differs from a coppicing system, by allowing revegetation by seedlings. Additionally, destructive forms of forest management are commonly referred to as 'clearcutting'.

However it has been argued that in New Zealand clearcutting is more environmentally friendly as it uses less resources for logging and less land for wood production allowing other land to be kept as old growth forest.

Clearcutting Regeneration, Harvesting or System

Clearcutting can be differentiated into

- *Clearcutting* - clean felling by complete exploitation and removal of all the trees in one operation ... a harvesting method
- *Clearcutting method* - method for regenerating an even-aged community by removing all the mature trees
- *Clearcutting system* - a silvicultural system incorporating the clearcutting method to remove (clear) the mature community over a considerable area at one time

Confusion between these different uses of the term is common. Furthermore as indicated above many variations mean technically correct usage may not be descriptive enough to know what is meant in that particular occasion.

Clearcutting in Southern Finland

Clearcutting near Eugene, Oregon

Effects on the Environment

Environmental groups criticize clear-cutting as destructive to water, soil, wildlife, and atmosphere, and recommend the use of sustainable alternatives. Clear-cutting has a very big impact on the water cycle. Trees hold water and topsoil. Clear-cutting in forests removes the trees which would otherwise have been transpiring large volumes of water and also physically damages the grasses, mosses, lichens, and ferns populating the understorey. All this bio-mass normally retains water during rainfall. Removal or damage of the biota reduces the local capacity to retain water, which can exacerbate flooding and lead to increased leaching of nutrients from the soil. The maximum nutrient loss occurs around year two, and returns to pre-clearcutting levels by year four.

Clear-cutting also prevents trees from shading riverbanks, which raises the temperature of riverbanks and rivers, contributing to the extinction of some fish and amphibian species. Because the trees no longer hold down the soil, riverbanks increasingly erode as sediment into the water, creating excess nutrients which exacerbate the changes in the river and create problems miles away, in the sea. All of the extra sediment and nutrients that leach into the streams cause the acidity of the stream to increase, which can kill marine life if the increase is great enough. The nutrient content of the soil was found to return to five percent of pre-clearcutting levels after 64 years, which demonstrates how clearcutting affects the environment for many years.

Clearcutting can destroy an area's ecological integrity in a number of ways, including: the destruction of buffer zones which reduce the severity of flooding by absorbing and holding water; the immediate removal of forest canopy, which destroys the habitat for many rainforest-dependent insects and bacteria; the removal of forest carbon sinks, leading to global warming through the increased human-induced and natural carbon dioxide build-up in the atmosphere; the elimination of fish and wildlife species due to soil erosion and habitat loss; the removal of underground worms, fungi and bacteria that condition soil and protect plants growing in it from disease; the loss of small-scale economic opportunities, such as fruit-picking, sap extraction, and rubber tapping; and the destruction of aesthetic values and recreational opportunities.

Negative Impacts

Clearcutting can have major negative impacts, both for humans and local flora and fauna. A study from the University of Oregon found that in certain zones, areas that were clear cut had nearly three times the amount of erosion due to slides. When the roads required by the clearcutting were factored in, the increase in slide activity appeared to be about 5 times greater compared to nearby forested areas. The roads built for clearcutting interrupt normal surface drainage because the roads are not as permeable as the normal ground cover. The roads also change subsurface water movement due to the redistribution of soil and rock. Clearcutting may lead to increased stream flow during storms, loss of habitat and species diversity, opportunities for invasive and weedy species, and negative impacts on scenery, as well as a decrease in property values; diminished recreation, hunting, and fishing opportunities. Clearcutting decreases the occurrence of natural disturbances like forest fires and natural uprooting. Over time, this can deplete the local seed bank. An example of what clearcutting did in Ontario before 1900 can be found in Edmund Zavitz.

In temperate and boreal climates, clearcutting can have an effect on the depth of snow, which is usually greater in a clearcut area than in the forest, due to a lack of interception and evapotranspiration. This results in less soil frost, which in combination with higher levels of direct sunlight results in snowmelt occurring earlier in the spring and earlier peak runoff.

The world's rain forests could completely vanish in a hundred years at the current rate of deforestation. Between June 2000 and June 2008 more than 150 000 square kilometers of rain forest were cleared in the Brazilian Amazon. Huge areas of forest have already been lost. For example, only eight to fourteen percent of the Atlantic Forest in South America now remains. While deforestation rates have slowed since 2004, forest loss is expected to continue for the foreseeable future. Farmers slash and burn large parcels of forest every year to create grazing and crop lands, but the forest's nutrient-poor soil often renders the land ill-suited for agriculture, and within a year or two, the farmers move on.

Positive Perspectives

Clearcutting can be practiced to encourage the growth and proliferation of tree species that require high light intensity. Generally, a harvest area wider than double the height of the adjacent trees will no longer be subject to the moderating influence of the woodland on the microclimate. The width of the harvest area can thus determine which species will come to dominate. Those with high tolerance to extremes in temperature, soil moisture, and resistance to browsing may be established, in particular secondary successional pioneer species.

Clearcutting can be used by foresters as a method of mimicking a natural disturbance and increasing primary successional species, such as poplar (aspen), willow and black cherry in North America. Clearcutting has also proved to be effective in creating animal habitat and browsing areas, which otherwise would not exist without natural stand-replacing disturbances such as wildfires, large scale windthrow, or avalanches.

Clearcuts are used to help regenerate species that cannot compete in mature forests. A number of them are aspen, jack pine and, in areas with poor soils, oaks—are important species for both game and nongame wildlife species. Clearcutting can also lead to increased vascular-plant diversity in the area. This is most pronounced after a couple years of clearcutting and in herb-rich forests where scarification took place.

No significant changes in water temperature were observed when patch clearcutting was done 100 feet away from a river. This suggests that patch clearcutting is a possible solution to concerns about changes in water temperature due to clearcutting. The effects of clearcutting on soil nutrient content were not examined in this study.

More recently, forest managers have found that clearcutting oak stands helps regenerate oak forests in areas of poor soil. The tree canopies in oak forests often shade out the ground, making it impossible for newly sprouted oaks to grow. When the mature trees are removed, the saplings stand a chance of recruiting into the forest.

Effects on Wildlife

Clearcutting's main destruction is towards habitats, where it makes the habitats more vulnerable in the future to damage by insects, diseases, acid rain, and wind. Removal of all trees from an area destroys the physical habitats of many species in wildlife. Also clearcutting can contribute to problems for ecosystems that depend on forests, like the streams and rivers which run through them.

In Canada, the black-tailed deer population is at further risk after clearcutting. The deer are a food source for wolves and cougars, as well as First Nations and other hunters. While deer may not be at risk in cities and rural countryside, where they can be seen running through neighbourhoods and feeding on farms, in higher altitude areas they require forest shelter.

Wildfire

A wildfire in California on September 5, 2008

A wildfire or wildland fire is a fire in an area of combustible vegetation that occurs in the countryside or rural area. Depending on the type of vegetation where it occurs, a wildfire can also be classified more specifically as a brush fire, bush fire, desert fire, forest fire, grass fire, hill fire, peat fire, vegetation fire, or veld fire. Fossil charcoal indicates that wildfires began soon after the appearance of terrestrial plants 420 million years ago. Wildfire's occurrence throughout the history of terrestrial life invites conjecture that fire must have had pronounced evolutionary effects on most ecosystems' flora and fauna. Earth is an intrinsically flammable planet owing to its cover of carbon-rich vegetation, seasonally dry climates, atmospheric oxygen, and wide-spread lightning and volcano ignitions.

Wildfires can be characterized in terms of the cause of ignition, their physical properties, the combustible material present, and the effect of weather on the fire. Wildfires can cause damage to property and human life, but they have many beneficial effects on native vegetation, animals, and ecosystems that have evolved with fire. Many plant species depend on the effects of fire for growth and reproduction. However, wildfire in ecosystems where wildfire is uncommon or where non-native vegetation has encroached may have negative ecological effects. Wildfire behaviour and severity result from the combination of factors such as available fuels, physical setting, and weather. Analyses of historical meteorological data and national fire records in western North America show the primacy of climate in driving large regional fires via wet periods that create substantial fuels or drought and warming that extend conducive fire weather.

Strategies of wildfire prevention, detection, and suppression have varied over the years. One common and inexpensive technique is controlled burning: permitting or even igniting smaller fires to minimize the amount of flammable material available for a potential wildfire. Vegetation may be burned periodically to maintain high species diversity and frequent burning of surface fuels limits fuel accumulation. Wildland fire use is the cheapest and most ecologically appropriate policy for many forests. Fuels may also be removed by logging, but fuels treatments and thinning have no effect on severe fire behavior Wildfire itself is reportedly "the most effective treatment for reducing a fire's rate of spread, fireline intensity, flame length, and heat per unit of area" according to Jan Van Wagtendonk, a biologist at the Yellowstone Field Station. Building codes in fire-prone areas typically require that structures be built of flame-resistant materials and a defensible space be maintained by clearing flammable materials within a prescribed distance from the structure.

Causes

Four major natural causes of wildfire ignitions exist:

- lightning
- sparks from rockfalls
- spontaneous combustion
- volcanic eruption

The most common direct human causes of wildfire ignition include arson, discarded cigarettes, power-line arcs (as detected by arc mapping), and sparks from equipment Ignition of wildland fires via contact with hot rifle-bullet fragments is also possible under the right conditions. Wildfires can also be started in communities experiencing shifting cultivation, where land is cleared quickly and farmed until the soil loses fertility, and slash and burn clearing. Forested areas cleared by logging encourage the dominance of flammable grasses, and abandoned logging roads overgrown by vegetation may act as fire corridors. Annual grassland fires in southern Vietnam stem in part from the destruction of forested areas by US military herbicides, explosives, and mechanical land-clearing and -burning operations during the Vietnam War.

The most common cause of wildfires varies throughout the world. In Canada and northwest China, for example, lightning operates as the major source of ignition. In other parts of the world, human involvement is a major contributor. In Africa, Central America, Fiji, Mexico, New Zealand, South America, and Southeast Asia, wildfires can be attributed to human activities such as agriculture, animal husbandry, and land-conversion burning. In China and in the Mediterranean Basin, human carelessness is a major cause of wildfires. In the United States and Australia, the source of wildfires can be traced both to lightning strikes and to human activities (such as machinery sparks, cast-away cigarette butts, or arson)." Coal seam fires burn in the thousands around the world, such as those in Burning Mountain, New South Wales; Centralia, Pennsylvania; and several coal-sustained fires in China. They can also flare up unexpectedly and ignite nearby flammable material.

Spread

A surface fire in the western desert of Utah, U.S.

Charred landscape following a crown fire in the North Cascades, U.S.

The spread of wildfires varies based on the flammable material present, its vertical arrangement and moisture content, and weather conditions. Fuel arrangement and density is governed in part by topography, as land shape determines factors such as available sunlight and water for plant growth. Overall, fire types can be generally characterized by their fuels as follows:

- Ground fires are fed by subterranean roots, duff and other buried organic matter. This fuel type is especially susceptible to ignition due to spotting. Ground fires typically burn by smoldering, and can burn slowly for days to months, such as peat fires in Kalimantan and Eastern Sumatra, Indonesia, which resulted from a riceland creation project that unintentionally drained and dried the peat.

- Crawling or surface fires are fueled by low-lying vegetation such as leaf and timber litter, debris, grass, and low-lying shrubbery.

- Ladder fires consume material between low-level vegetation and tree canopies, such as small trees, downed logs, and vines. Kudzu, Old World climbing fern, and other invasive plants that scale trees may also encourage ladder fires.

- Crown, canopy, or aerial fires burn suspended material at the canopy level, such as tall trees, vines, and mosses. The ignition of a crown fire, termed *crowning*, is dependent on the density of the suspended material, canopy height, canopy continuity, sufficient surface and ladder fires, vegetation moisture content, and weather conditions during the blaze. Stand-replacing fires lit by humans can spread into the Amazon rain forest, damaging ecosystems not particularly suited for heat or arid conditions.

Physical Properties

Experimental fire in Canada

Wildfires occur when all of the necessary elements of a fire triangle come together in a susceptible area: an ignition source is brought into contact with a combustible material such as vegetation, that is subjected to sufficient heat and has an adequate supply of oxygen from the ambient air. A high moisture content usually prevents ignition and slows propagation, because higher temperatures are required to evaporate any water within the material and heat the material to its fire point. Dense forests usually provide more shade, resulting in lower ambient temperatures and greater humidity, and are therefore less susceptible to wildfires. Less dense material such as grasses and leaves are easier to ignite because they contain less water than denser material such as branches and trunks. Plants continuously lose water by evapotranspiration, but water loss is usually balanced by water absorbed from the soil, humidity, or rain. When this balance is not maintained, plants dry out and are therefore more flammable, often a consequence of droughts.

A wildfire *front* is the portion sustaining continuous flaming combustion, where unburned material meets active flames, or the smoldering transition between unburned and burned material. As the front approaches, the fire heats both the surrounding air and woody material through convection and thermal radiation. First, wood is dried as water is vaporized at a temperature of 100 °C (212 °F). Next, the pyrolysis of wood at 230 °C (450 °F) releases flammable gases. Finally, wood can smoulder at 380 °C (720 °F) or, when heated sufficiently, ignite at 590 °C (1,000 °F). Even before the flames of a wildfire arrive at a particular location, heat transfer from the wildfire front warms the air to 800 °C (1,470 °F), which pre-heats and dries flammable materials, causing materials to ignite faster and allowing the fire to spread faster. High-temperature and long-duration surface wildfires may encourage flashover or *torching*: the drying of tree canopies and their subsequent ignition from below.

Torching of Juniper Tree on the Palisade Wildfire in Nevada

Wildfires have a rapid *forward rate of spread* (FROS) when burning through dense, uninterrupted fuels. They can move as fast as 10.8 kilometres per hour (6.7 mph) in forests and 22 kilometres per hour (14 mph) in grasslands. Wildfires can advance tangential to the main front to form a *flanking* front, or burn in the opposite direction of the main front by *backing*. They may also spread by *jumping* or *spotting* as winds and vertical convection columns carry *firebrands* (hot wood embers) and other burning materials through the air over roads, rivers, and other barriers that may otherwise act as firebreaks. Torching and fires in tree canopies encourage spotting, and dry ground fuels that surround a wildfire are especially vulnerable to ignition from firebrands. Spotting can create *spot fires* as hot embers and firebrands ignite fuels downwind from the fire. In Australian bushfires, spot fires are known to occur as far as 20 kilometres (12 mi) from the fire front.

Especially large wildfires may affect air currents in their immediate vicinities by the stack effect: air rises as it is heated, and large wildfires create powerful updrafts that will draw in new, cooler air from surrounding areas in thermal columns. Great vertical differences in temperature and humidity encourage pyrocumulus clouds, strong winds, and fire whirls with the force of tornadoes at speeds of more than 80 kilometres per hour (50 mph). Rapid rates of spread, prolific crowning or spotting, the presence of fire whirls, and strong convection columns signify extreme conditions.

The thermal heat from wildfire can cause significant weathering of rocks and boulders, heat can rapidly expand a boulder and thermal shock can occur, which may cause an object's structure to fail.

Effect of Weather

Lightning-sparked wildfires are frequent occurrences during the dry summer season in Nevada.

A wildfire in Venezuela during a drought

Heat waves, droughts, cyclical climate changes such as El Niño, and regional weather patterns such as high-pressure ridges can increase the risk and alter the behavior of wildfires dramatically. Years of precipitation followed by warm periods can encourage more widespread fires and longer fire seasons. Since the mid-1980s, earlier snowmelt and associated warming has also been associated with an increase in length and severity of the wildfire season in the Western United States. Global warming may increase the intensity and frequency of droughts in many areas, creating more intense and frequent wildfires. A 2015 study indicates that the increase in fire risk in California may be attributable to human-induced climate change. A study of alluvial sediment deposits going back over 8,000 years found warmer climate periods experienced severe droughts and stand-replacing fires and concluded climate was such a powerful influence on wildfire that trying to recreate presettlement forest structure is likely impossible in a warmer future.

Intensity also increases during daytime hours. Burn rates of smoldering logs are up to five times greater during the day due to lower humidity, increased temperatures, and increased wind speeds. Sunlight warms the ground during the day which creates air currents that travel uphill. At night the land cools, creating air currents that travel downhill. Wildfires are fanned by these winds and often follow the air currents over hills and through valleys. Fires in Europe occur frequently during the hours of 12:00 p.m. and 2:00 p.m. Wildfire suppression operations in the United States revolve around a 24-hour *fire day* that begins at 10:00 a.m. due to the predictable increase in intensity resulting from the daytime warmth.

Ecology

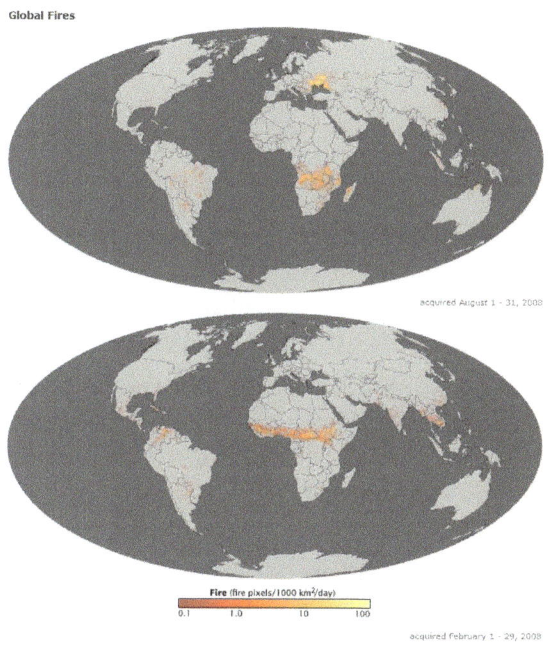

Global fires during the year 2008 for the months of August (top image) and February (bottom image), as detected by the Moderate Resolution Imaging Spectroradiometer (MODIS) on NASA's Terra satellite.

Wildfire's occurrence throughout the history of terrestrial life invites conjecture that fire must have had pronounced evolutionary effects on most ecosystems' flora and fauna. Wildfires are common in climates that are sufficiently moist to allow the growth of vegetation but feature extended dry, hot periods. Such places include the vegetated areas of Australia and Southeast Asia, the veld in southern Africa, the fynbos in the Western Cape of South Africa, the forested areas of the United States and Canada, and the Mediterranean Basin.

High-severity wildfire creates complex early seral habitat (also called "snag forest habitat"), which often has higher species richness and diversity than unburned old forest. Plant and animal species in most types of North American forests evolved with fire, and many of these species depend on wildfires, and particularly high-severity fires, to reproduce and grow. Fire helps to return nutrients from plant matter back to soil, the heat from fire is necessary to the germination of certain types of seeds, and the snags (dead trees) and early successional forests created by high-severity fire create habitat conditions that are beneficial to wildlife. Early successional forests created by high-severity fire support some of the highest levels of native biodiversity found in temperate conifer forests. Post-fire logging has no ecological benefits and many negative impacts; the same is often true for post-fire seeding.

Although some ecosystems rely on naturally occurring fires to regulate growth, some ecosystems suffer from too much fire, such as the chaparral in southern California and lower elevation deserts in the American Southwest. The increased fire frequency in these ordinarily fire-dependent areas has upset natural cycles, damaged native plant communities, and encouraged the growth of non-native weeds. Invasive species, such as *Lygodium microphyllum* and *Bromus tectorum*, can grow rapidly in areas that were damaged by fires. Because they are highly flammable, they can increase the future risk of fire, creating a positive feedback loop that increases fire frequency and further alters native vegetation communities.

In the Amazon Rainforest, drought, logging, cattle ranching practices, and slash-and-burn agriculture damage fire-resistant forests and promote the growth of flammable brush, creating a cycle that encourages more burning. Fires in the rainforest threaten its collection of diverse species and produce large amounts of CO_2. Also, fires in the rainforest, along with drought and human involvement, could damage or destroy more than half of the Amazon rainforest by the year 2030. Wildfires generate ash, destroy available organic nutrients, and cause an increase in water runoff, eroding away other nutrients and creating flash flood conditions. A 2003 wildfire in the North Yorkshire Moors destroyed 2.5 square kilometers (600 acres) of heather and the underlying peat layers. Afterwards, wind erosion stripped the ash and the exposed soil, revealing archaeological remains dating back to 10,000 BC. Wildfires can also have an effect on climate change, increasing the amount of carbon released into the atmosphere and inhibiting vegetation growth, which affects overall carbon uptake by plants.

In tundra there is a natural pattern of accumulation of fuel and wildfire which varies depending on the nature of vegetation and terrain. Research in Alaska has shown fire-event return intervals, (FRIs) that typically vary from 150 to 200 years with dryer lowland areas burning more frequently than wetter upland areas.

Plant Adaptation

Ecological succession after a wildfire in a boreal pine forest next to Hara Bog, Lahemaa National Park, Estonia. The pictures were taken one and two years after the fire.

Plants in wildfire-prone ecosystems often survive through adaptations to their local fire regime. Such adaptations include physical protection against heat, increased growth after a fire event, and flammable materials that encourage fire and may eliminate competition. For example, plants of the genus *Eucalyptus* contain flammable oils that encourage fire and hard sclerophyll leaves to resist heat and drought, ensuring their dominance over less fire-tolerant species. Dense bark, shedding lower branches, and high water content in external structures may also protect trees from rising temperatures. Fire-resistant seeds and reserve shoots that sprout after a fire encourage species preservation, as embodied by pioneer species. Smoke, charred wood, and heat can stimulate the germination of seeds in a process called *serotiny*. Exposure to smoke from burning plants promotes germination in other types of plants by inducing the production of the orange butenolide.

Grasslands in Western Sabah, Malaysian pine forests, and Indonesian *Casuarina* forests are believed to have resulted from previous periods of fire. Chamise deadwood litter is low in water content and flammable, and the shrub quickly sprouts after a fire. Cape lilies lie dormant until flames brush away the covering, then blossom almost overnight. Sequoia rely on periodic fires to reduce competition, release seeds from their cones, and clear the soil and canopy for new growth. Caribbean Pine in Bahamian pineyards have adapted to and rely on low-intensity, surface fires for survival and growth. An optimum fire frequency for growth is every 3 to 10 years. Too frequent fires favor herbaceous plants, and infrequent fires favor species typical of Bahamian dry forests.

Atmospheric Effects

A Pyrocumulus cloud produced by a wildfire in Yellowstone National Park

Most of the Earth's weather and air pollution resides in the troposphere, the part of the atmosphere that extends from the surface of the planet to a height of about 10 kilometers (6 mi). The vertical lift of a severe thunderstorm or pyrocumulonimbus can be enhanced in the area of a large wildfire, which can propel smoke, soot, and other particulate matter as high as the lower stratosphere. Previously, prevailing scientific theory held that most particles in the stratosphere came from volcanoes, but smoke and other wildfire emissions have been detected from the lower stratosphere. Pyrocumulus clouds can reach 6,100 meters (20,000 ft) over wildfires. Satellite observation of smoke plumes from wildfires revealed that the plumes could be traced intact for distances exceeding 1,600 kilometers (1,000 mi). Computer-aided models such as CALPUFF may help predict the size and direction of wildfire-generated smoke plumes by using atmospheric dispersion modeling.

Wildfires can affect local atmospheric pollution, and release carbon in the form of carbon dioxide. Wildfire emissions contain fine particulate matter which can cause cardiovascular and respiratory problems. Increased fire byproducts in the troposphere can increase ozone concentration beyond safe levels. Forest fires in Indonesia in 1997 were estimated to have released between 0.81 and 2.57 gigatonnes (0.89 and 2.83 billion short tons) of CO_2 into the atmosphere, which is between 13%–40% of the annual global carbon dioxide emissions from burning fossil fuels. Atmospheric models suggest that these concentrations of sooty particles could increase absorption of incoming solar radiation during winter months by as much as 15%.

Smoke trail from a fire seen while looking towards Dargo from Swifts Creek, Victoria, Australia, 11 January 2007

History

In the Welsh Borders, the first evidence of wildfire is rhyniophytoid plant fossils preserved as charcoal, dating to the Silurian period (about 420 million years ago). Smoldering surface fires started to occur sometime before the Early Devonian period 405 million years ago. Low atmospheric oxygen during the Middle and Late Devonian was accompanied by a decrease in charcoal abundance. Additional charcoal evidence suggests that fires continued through the Carboniferous period. Later, the overall increase of atmospheric oxygen from 13% in the Late Devonian to 30-31% by the Late Permian was accompanied by a more widespread distribution of wildfires. Later, a decrease in wildfire-related charcoal deposits from the late Permian to the Triassic periods is explained by a decrease in oxygen levels.

Wildfires during the Paleozoic and Mesozoic periods followed patterns similar to fires that occur in modern times. Surface fires driven by dry seasons are evident in Devonian and Carboniferous progymnosperm forests. Lepidodendron forests dating to the Carboniferous period have charred peaks, evidence of crown fires. In Jurassic gymnosperm forests, there is evidence of high frequency, light surface fires. The increase of fire activity in the late Tertiary is possibly due to the increase of C_4-type grasses. As these grasses shifted to more mesic habitats, their high flammability increased fire frequency, promoting grasslands over woodlands. However, fire-prone habitats may have contributed to the prominence of trees such as those of the genera *Eucalyptus*, *Pinus* and *Sequoia*, which have thick bark to withstand fires and employ serotiny.

Human Involvement

Aerial view of deliberate wildfires on the Khun Tan Range, Thailand. These fires are lit by local farmers every year in order to promote the growth of a certain mushroom

The human use of fire for agricultural and hunting purposes during the Paleolithic and Mesolithic ages altered the preexisting landscapes and fire regimes. Woodlands were gradually replaced by smaller vegetation that facilitated travel, hunting, seed-gathering and planting. In recorded human history, minor allusions to wildfires were mentioned in the Bible and by classical writers such as Homer. However, while ancient Hebrew, Greek, and Roman writers were aware of fires, they were not very interested in the uncultivated lands where wildfires occurred. Wildfires were used in battles throughout human history as early thermal weapons. From the Middle ages, accounts were written of occupational burning as well as customs and laws that governed the use of fire. In Germany, regular burning was documented in 1290 in the Odenwald and in 1344 in the Black Forest. In the 14th century Sardinia, firebreaks were used for wildfire protection. In Spain during the 1550s, sheep husbandry was discouraged in certain provinces by Philip II due to the harmful effects of fires used in transhumance. As early as the 17th century, Native Americans were observed using fire for many purposes including cultivation, signaling, and warfare. Scottish botanist David Douglas noted the native use of fire for tobacco cultivation, to encourage deer into smaller areas for hunting purposes, and to improve foraging for honey and grasshoppers. Charcoal found in sedimentary deposits off the Pacific coast of Central America suggests that more burning occurred in the 50 years before the Spanish colonization of the Americas than after the colonization. In the post-World War II Baltic region, socio-economic changes led more stringent air quality standards and bans on fires that eliminated traditional burning practices. In the mid-19th century, explorers from the HMS *Beagle* observed Australian Aborigines using fire for ground clearing, hunting, and regeneration of plant food in a method later named fire-stick farming.

Such careful use of fire has been employed for centuries in the lands protected by Kakadu National Park to encourage biodiversity.

Wildfires typically occurred during periods of increased temperature and drought. An increase in fire-related debris flow in alluvial fans of northeastern Yellowstone National Park was linked to the period between AD 1050 and 1200, coinciding with the Medieval Warm Period. However, human influence caused an increase in fire frequency. Dendrochronological fire scar data and charcoal layer data in Finland suggests that, while many fires occurred during severe drought conditions, an increase in the number of fires during 850 BC and 1660 AD can be attributed to human influence. Charcoal evidence from the Americas suggested a general decrease in wildfires between 1 AD and 1750 compared to previous years. However, a period of increased fire frequency between 1750 and 1870 was suggested by charcoal data from North America and Asia, attributed to human population growth and influences such as land clearing practices. This period was followed by an overall decrease in burning in the 20th century, linked to the expansion of agriculture, increased livestock grazing, and fire prevention efforts. A meta-analysis found that 17 times more land burned annually in California before 1800 compared to recent decades (1,800,000 hectares/year compared to 102,000 hectares/year).

Invasive species moved by humans have in some cases increased the intensity of wildfires, such as Eucalyptus in California and gamba grass in Australia.

Prevention

1985 Smokey Bear poster with part of his admonition, "Only you can prevent forest fires".

Wildfire prevention refers to the preemptive methods aimed at reducing the risk of fires as well as lessening its severity and spread. Prevention techniques aim to manage air quality, maintain ecological balances, protect resources, and to affect future fires. North American firefighting policies permit naturally caused fires to burn to maintain their ecological role, so long as the risks of escape into high-value areas are mitigated. However, prevention policies must consider the role that humans play in wildfires, since, for example, 95% of forest fires in Europe are related to human involvement. Sources of human-caused fire may include arson, accidental ignition, or the uncontrolled use of fire in land-clearing and agriculture such as the slash-and-burn farming in Southeast Asia.

In 1937, U.S. President Franklin D. Roosevelt initiated a nationwide fire prevention campaign, highlighting the role of human carelessness in forest fires. Later posters of the program featured Uncle Sam, leaders of the Axis powers of World War II, characters from the Disney movie *Bambi*, and the official mascot of the U.S. Forest Service, Smokey Bear. Reducing human-caused ignitions may be the most effective means of reducing unwanted wildfire. Alteration of fuels is commonly undertaken when attempting to affect future fire risk and behavior. Wildfire prevention programs around the world may employ techniques such as *wildland fire use* and *prescribed or controlled burns*. *Wildland fire use* refers to any fire of natural causes that is monitored but allowed to burn. *Controlled burns* are fires ignited by government agencies under less dangerous weather conditions.

A prescribed burn in a *Pinus nigra* stand in Portugal

Vegetation may be burned periodically to maintain high species diversity and frequent burning of surface fuels limits fuel accumulation. Wildland fire use is the cheapest and most ecologically appropriate policy for many forests. Fuels may also be removed by logging, but fuels treatments and thinning have no effect on severe fire behavior Wildfire models are often used to predict and compare the benefits of different fuel treatments on future wildfire spread, but their accuracy is low.

Wildfire itself is reportedly "the most effective treatment for reducing a fire's rate of spread, fireline intensity, flame length, and heat per unit of area" according to Jan Van Wagtendonk, a biologist at the Yellowstone Field Station.

Building codes in fire-prone areas typically require that structures be built of flame-resistant materials and a defensible space be maintained by clearing flammable materials within a prescribed distance from the structure. Communities in the Philippines also maintain fire lines 5 to 10 meters (16 to 33 ft) wide between the forest and their village, and patrol these lines during summer months or seasons of dry weather. Continued residential development in fire-prone areas and rebuilding structures destroyed by fires has been met with criticism. The ecological benefits of fire are often overridden by the economic and safety benefits of protecting structures and human life.

Us Wildfire Policy

Poster for forest fire prevention showing a burning cigarette and a forest fire.

History of Wildfire Policy in the U.S.

Since the turn of the 20th century, various federal and state agencies have been involved in wildland fire management in one form or another. In the early 20th century, for example, the federal government, through the U.S. Army and the U.S. Forest Service, solicited fire suppression as a primary goal of managing the nation's forests. At this time in history fire was viewed as a threat to timber, an economically important natural resource. As such, the decision was made to devote public funds to fire suppression and fire prevention efforts. For example, the Forest Fire Emergency Fund Act of 1908 permitted deficit spending in the case of emergency fire situations. As a result, the U.S. Forest Service was able to acquire a deficit of over $1 million in 1910 due to emergency fire suppression efforts. Following the same tone of timber resource protection, the U.S. Forest Service adopted the "10 AM Policy" in 1935. Through this policy, the agency advocated the control of all fires by 10 o'clock of the morning following the discovery of a wildfire. Fire prevention was also heavily advocated through public education campaigns such as Smokey Bear. Through these and similar public education campaigns the general public was, in a sense, trained to perceive all wildfire as a threat to civilized society and natural resources. The negative sentiment towards wildland fire prevailed and helped to shape wildland fire management objectives throughout most of the 20th century.

Texas was particularly hard-hit by wildfires in 2011, as noted by this placard at the state Forestry Museum in Lufkin.

Beginning in the 1970s public perception of wildland fire management began to shift. Despite strong funding for fire suppression in the first half of the 20th century, massive wildfires continued to be prevalent across the landscape of North America. Ecologists were beginning to recognize the presence and ecological importance of natural, lightning-ignited wildfires across the United States. It was learned that suppression of fire in certain ecosystems may in fact increase the likelihood that a wildfire will occur and may increase the intensity of those wildfires. With the emergence of fire ecology as a science also came an effort to apply fire to ecosystems in a controlled manner; however, suppression is still the main tactic when a fire is set by a human or if it threatens life or property. By the 1980s, in light of this new understanding, funding efforts began to support prescribed burning in order to prevent wildfire events. In 2001, the United States implemented a National Fire Plan, increasing the budget for the reduction of hazardous fuels from $108 million in 2000 to $401 million.

In addition to using prescribed fire to reduce the chance of catastrophic wildfires, mechanical methods have recently been adopted as well. Mechanical methods include the use of chippers and other machinery to remove hazardous fuels and thereby reduce the risk of wildfire events. Today the United States' maintains that, "fire, as a critical natural process, will be integrated into land and resource management plans and activities on a landscape scale, and across agency boundaries. Response to wildfire is based on ecological, social and legal consequences of fire. The circumstance under which a fire occurs, and the likely consequences and public safety and welfare, natural and cultural resources, and values to be protected dictate the appropriate management response to fire" (United States Department of Agriculture Guidance for Implementation of Federal Wildland Fire Management Policy, 13 February 2009). The five federal regulatory agencies managing forest fire response and planning for 676 million acres in the United States are the Department of the Interior, the Bureau of Land Management, the Bureau of Indian Affairs, the National Park Service, the United States Department of Agriculture-Forest Service and the United States Fish and Wildlife Services. Several hundred million U.S. acres of wildfire management are also conducted by state, county, and local fire management organizations. In 2014, legislators proposed The Wildfire Disaster Funding Act to provide $2.7 billion fund appropriated by congress for the USDA and Department of Interior to use in fire suppression. The bill is a reaction to United States Forest Service and Department of Interior costs of Western Wildfire suppression appending that amounted to $3.5 billion in 2013.

Wildland-urban Interface Policy

An aspect of wildfire policy that is gaining attention is the wildland-urban interface (WUI). More and more people are living in "red zones," or areas that are at high risk of wildfires. FEMA and the NFPA develop specific policies to guide homeowners and builders in how to build and maintain structures at the WUI and how protect against property losses. For example, NFPA-1141 is a standard for fire protection infrastructure for land development in wildland, rural and suburban areas and NFPA-1144 is a standard for reducing structure ignition hazards from wildland fire. Compensation for losses in the WUI are typically negotiated on an incident-by-incident basis. This is generating discussion about the burden of responsibility for funding and fighting a fire in the WUI, in that, if a resident chooses to live in a known red zone, should he or she retain a higher level of responsibility for funding home protection against wildfires. One initiative aimed at helping U.S. WUI communities live more safely with fire is called fire-adapted communities.

Economics of Fire Management Policy

Similar to that of military operations, fire management is often very expensive in the U.S. and the rest of the world. Today, it is not uncommon for suppression operations for a single wildfire to exceed costs of $1 million in just a few days. The United States Department of Agriculture allotted $2.2 billion for wildfire management in 2012. Although fire suppression purports to benefit society, other options for fire management exist. While these options cannot completely replace fire suppression as a fire management tool, other options can play an important role in overall fire management and can therefore affect the costs of fire suppression.

It is commonly accepted that past fire suppression and climate change has resulted in larger, more intense wildfire events which are seen today. In economic terms, expenditures used for wildfire suppression in the early 20th century have contributed to increased suppression costs which are being realized today.

Detection

Fast and effective detection is a key factor in wildfire fighting. Early detection efforts were focused on early response, accurate results in both daytime and nighttime, and the ability to prioritize fire danger. Fire lookout towers were used in the United States in the early 20th century and fires were reported using telephones, carrier pigeons, and heliographs. Aerial and land photography using instant cameras were used in the 1950s until infrared scanning was developed for fire detection in the 1960s. However, information analysis and delivery was often delayed by limitations in communication technology. Early satellite-derived fire analyses were hand-drawn on maps at a remote site and sent via overnight mail to the fire manager. During the Yellowstone fires of 1988, a data station was established in West Yellowstone, permitting the delivery of satellite-based fire information in approximately four hours.

Dry Mountain Fire Lookout in the Ochoco National Forest, Oregon, circa 1930

Currently, public hotlines, fire lookouts in towers, and ground and aerial patrols can be used as a means of early detection of forest fires. However, accurate human observation may be limited by operator fatigue, time of day, time of year, and geographic location. Electronic systems have gained popularity in recent years as a possible resolution to human operator error. A government report on a recent trial of three automated camera fire detection systems in Australia did, however, conclude "...detection by the camera systems was slower and less reliable than by a trained human observer". These systems may be semi- or fully automated and employ systems based on the risk area and degree of human presence, as suggested by GIS data analyses. An integrated approach of multiple systems can be used to merge satellite data, aerial imagery, and personnel position via Global Positioning System (GPS) into a collective whole for near-realtime use by wireless Incident Command Centers.

A small, high risk area that features thick vegetation, a strong human presence, or is close to a critical urban area can be monitored using a local sensor network. Detection systems may include wireless sensor networks that act as automated weather systems: detecting temperature, humidity, and smoke. These may be battery-powered, solar-powered, or *tree-rechargeable*: able to recharge their battery systems using the small electrical currents in plant material. Larger, medium-risk areas can be monitored by scanning towers that incorporate fixed cameras and sensors to detect smoke or additional factors such as the infrared signature of carbon dioxide produced by fires. Additional capabilities such as night vision, brightness detection, and color change detection may also be incorporated into sensor arrays.

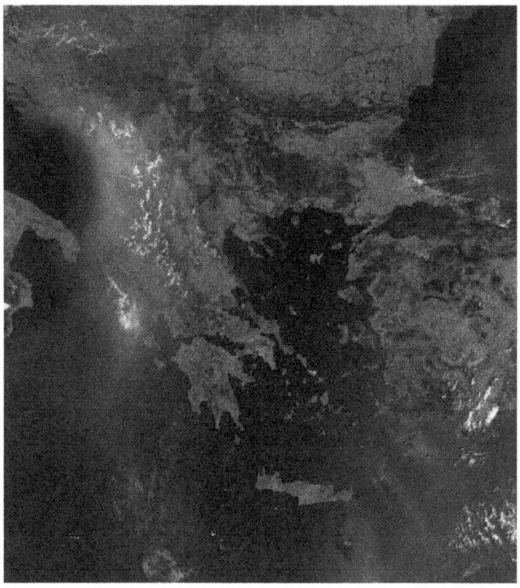

Wildfires across the Balkans in late July 2007 (MODIS image)

Satellite and aerial monitoring through the use of planes, helicopter, or UAVs can provide a wider view and may be sufficient to monitor very large, low risk areas. These more sophisticated systems employ GPS and aircraft-mounted infrared or high-resolution visible cameras to identify and target wildfires. Satellite-mounted sensors such as Envisat's Advanced Along Track Scanning Radiometer and European Remote-Sensing Satellite's Along-Track Scanning Radiometer can measure infrared radiation emitted by fires, identifying hot spots greater than 39 °C (102 °F). The National Oceanic and Atmospheric Administration's Hazard Mapping System combines remote-sensing data from satellite sources such as Geostationary Operational Environmental Satellite (GOES), Moderate-Resolution Imaging Spectroradiometer (MODIS), and Advanced Very High Resolution Radiometer (AVHRR) for detection of fire and smoke plume locations. However, satellite detection is prone to offset errors, anywhere from 2 to 3 kilometers (1 to 2 mi) for MODIS and AVHRR data and up to 12 kilometers (7.5 mi) for GOES data. Satellites in geostationary orbits may become disabled, and satellites in polar orbits are often limited by their short window of observation time. Cloud cover and image resolution and may also limit the effectiveness of satellite imagery.

in 2015 a new fire detection tool is in operation at the U.S. Department of Agriculture (USDA) Forest Service (USFS) which uses data from the Suomi National Polar-orbiting Partnership (NPP) satellite to detect smaller fires in more detail than previous space-based products. The high-resolution data is used with a computer model to predict how a fire will change direction based on weather and land conditions. The active fire detection product using data from Suomi NPP's Visible Infrared Imaging

Radiometer Suite (VIIRS) increases the resolution of fire observations to 1,230 feet (375 meters). Previous NASA satellite data products available since the early 2000s observed fires at 3,280 foot (1 kilometer) resolution. The data is one of the intelligence tools used by the USFS and Department of Interior agencies across the United States to guide resource allocation and strategic fire management decisions. The enhanced VIIRS fire product enables detection every 12 hours or less of much smaller fires and provides more detail and consistent tracking of fire lines during long duration wildfires – capabilities critical for early warning systems and support of routine mapping of fire progression. Active fire locations are available to users within minutes from the satellite overpass through data processing facilities at the USFS Remote Sensing Applications Center, which uses technologies developed by the NASA Goddard Space Flight Center Direct Readout Laboratory in Greenbelt, Maryland. The model uses data on weather conditions and the land surrounding an active fire to predict 12–18 hours in advance whether a blaze will shift direction. The state of Colorado decided to incorporate the weather-fire model in its firefighting efforts beginning with the 2016 fire season.

In 2014, an international campaign was organized in South Africa's Kruger National Park to validate fire detection products including the new VIIRS active fire data. In advance of that campaign, the Meraka Institute of the Council for Scientific and Industrial Research in Pretoria, South Africa, an early adopter of the VIIRS 375m fire product, put it to use during several large wildfires in Kruger.

The demand for timely, high-quality fire information has increased in recent years. Wildfires in the United States burn an average of 7 million acres of land each year. For the last 10 years, the USFS and Department of Interior have spent a combined average of about $2–4 billion annually on wildfire suppression.

Suppression

A Russian firefighter extinguishing a wildfire

Wildfire suppression depends on the technologies available in the area in which the wildfire occurs. In less developed nations the techniques used can be as simple as throwing sand or beating the fire with sticks or palm fronds. In more advanced nations, the suppression methods vary due to increased technological capacity. Silver iodide can be used to encourage snow fall, while fire retardants and water can be dropped onto fires by unmanned aerial vehicles, planes, and helicopters. Complete fire suppression is no longer an expectation, but the majority of wildfires are often extinguished before they grow out of control. While more than 99% of the 10,000 new wildfires each year are contained, escaped wildfires under extreme weather conditions are difficult to suppress without a change in the weather. Wildfires in Canada and the US burn an average of 54,500 square kilometers (13,000,000 acres) per year.

Above all, fighting wildfires can become deadly. A wildfire's burning front may also change direction unexpectedly and jump across fire breaks. Intense heat and smoke can lead to disorientation and loss of appreciation of the direction of the fire, which can make fires particularly dangerous. For example, during the 1949 Mann Gulch fire in Montana, USA, thirteen smokejumpers died when they lost their communication links, became disoriented, and were overtaken by the fire. In the Australian February 2009 Victorian bushfires, at least 173 people died and over 2,029 homes and 3,500 structures were lost when they became engulfed by wildfire.

Costs of Wildfire Suppression

In California, the U.S. Forest Service spends about $200 million per year to suppress 98% of wildfires and up to $1 billion to suppress the other 2% of fires that escape initial attack and become large.

Wildland Firefighting Safety

Wildfire fighters cutting down a tree using a chainsaw

Wildland fire fighters face several life-threatening hazards including heat stress, fatigue, smoke and dust, as well as the risk of other injuries such as burns, cuts and scrapes, animal bites, and even rhabdomyolysis.

Especially in hot weather condition, fires present the risk of heat stress, which can entail feeling heat, fatigue, weakness, vertigo, headache, or nausea. Heat stress can progress into heat strain, which entails physiological changes such as increased heart rate and core body temperature. This can lead to heat-related illnesses, such as heat rash, cramps, exhaustion or heat stroke. Various factors can contribute to the risks posed by heat stress, including strenuous work, personal risk factors such as age and fitness, dehydration, sleep deprivation, and burdensome personal protective equipment. Rest, cool water, and occasional breaks are crucial to mitigating the effects of heat stress.

Smoke, ash, and debris can also pose serious respiratory hazards to wildland fire fighters. The smoke and dust from wildfires can contain gases such as carbon monoxide, sulfur dioxide and formaldehyde, as well as particulates such as ash and silica. To reduce smoke exposure, wildfire fighting crews should, whenever possible, rotate firefighters through areas of heavy smoke, avoid downwind firefighting, use equipment rather than people in holding areas, and minimize mop-up. Camps and command posts should also be located upwind of wildfires. Protective clothing and equipment can also help minimize exposure to smoke and ash.

Firefighters are also at risk of cardiac events including strokes and heart attacks. Fire fighters should maintain good physical fitness. Fitness programs, medical screening and examination programs which include stress tests can minimize the risks of firefighting cardiac problems. Other injury hazards wildland fire fighters face include slips, trips and falls, burns, scrapes and cuts from tools and equipment, being struck by trees, vehicles, or other objects, plant hazards such as thorns and poison ivy, snake and animal bites, vehicle crashes, electrocution from power lines or lightning storms, and unstable building structures.

Fire Retardant

Fire retardants are used to help slow wildfires, coat fuels, and lessen oxygen availability as required by various firefighting situations. They are composed of nitrates, ammonia, phosphates and sulfates, as well as other chemicals and thickening agents. The choice of whether to apply retardant depends on the magnitude, location and intensity of the wildfire. Fire retardants are used to reach inaccessible geographical regions where ground firefighting crews are unable to reach a wildfire or in any occasion where human safety and structures are endangered. In certain instances, fire retardant may also be applied ahead of wildfires for protection of structures and vegetation as a precautionary fire defense measure.

The application of aerial fire retardants creates an atypical appearance on land and water surfaces and has the potential to change soil chemistry. Fire retardant can decrease the availability of plant nutrients in the soil by increasing the acidity of the soil. Fire retardant may also affect water quality through leaching, eutrophication, or misapplication. Fire retardant's effects on drinking water remain inconclusive. Dilution factors, including water body size, rainfall, and water flow rates lessen the concentration and potency of fire retardant. Wildfire debris (ash and sediment) clog rivers and reservoirs increasing the risk for floods and erosion that ultimately slow and/or damage water treatment systems. There is continued concern of fire retardant effects on land, water, wildlife habitats, and watershed quality, additional research is needed. However, on the positive side, fire retardant (specifically its nitrogen and phosphorus components) has been shown to have a fertilizing effect on nutrient-deprived soils and thus creates a temporary increase in vegetation.

Current USDA procedure maintains that the aerial application of fire retardant in the United States must clear waterways by a minimum of 300 feet in order to safeguard effects of retardant runoff. Aerial uses of fire retardant are required to avoid application near waterways and endangered species (plant and animal habitats). After any incident of fire retardant misapplication, the U.S. Forest Service requires reporting and assessment impacts be made in order to determine mitigation, remediation, and/or restrictions on future retardant uses in that area.

Modeling

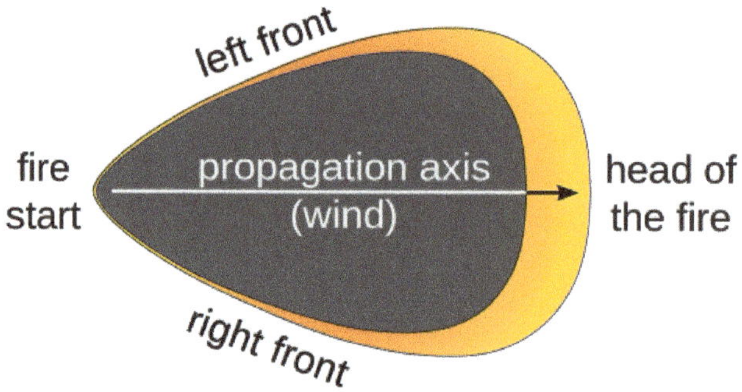

Fire Propagation Model

Wildfire modeling is concerned with numerical simulation of wildfires in order to comprehend and predict fire behavior. Wildfire modeling aims to aid wildfire suppression, increase the safety of firefighters and the public, and minimize damage. Using computational science, wildfire modeling involves the statistical analysis of past fire events to predict spotting risks and front behavior. Various wildfire propagation models have been proposed in the past, including simple ellipses and egg- and fan-shaped models. Early attempts to determine wildfire behavior assumed terrain and vegetation uniformity. However, the exact behavior of a wildfire's front is dependent on a variety of factors, including windspeed and slope steepness. Modern growth models utilize a combination of past ellipsoidal descriptions and Huygens' Principle to simulate fire growth as a continuously expanding polygon. Extreme value theory may also be used to predict the size of large wildfires. However, large fires that exceed suppression capabilities are often regarded as statistical outliers in standard analyses, even though fire policies are more influenced by large wildfires than by small fires.

Human Risk and Exposure

2009 California Wildfires at NASA/JPL – Pasadena, California

Wildfire risk is the chance that a wildfire will start in or reach a particular area and the potential loss of human values if it does. Risk is dependent on variable factors such as human activities, weather patterns, availability of wildfire fuels, and the availability or lack of resources to suppress a fire. Wildfires have continually been a threat to human populations. However, human induced geographical and climatic changes are exposing populations more frequently to wildfires and increasing wildfire risk. It is speculated that the increase in wildfires arises from a century of wildfire suppression coupled with the rapid expansion of human developments into fire-prone wildlands. Wildfires are naturally occurring events that aid in promoting forest health. Global warming and climate changes are causing an increase in temperatures and more droughts nationwide which contributes to an increase in wildfire risk.

Regional Burden of Wildfires in the United States

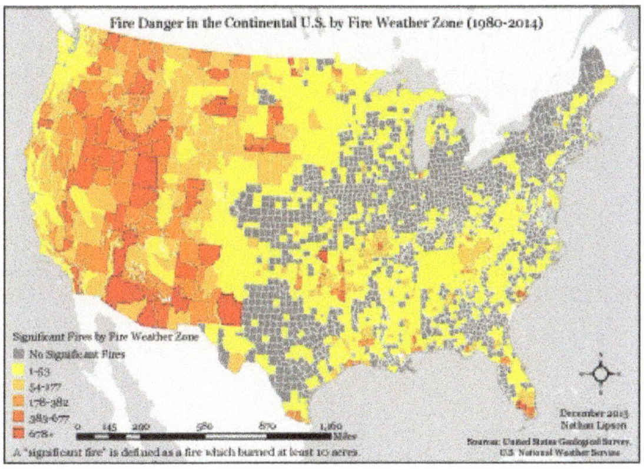

This map depicts regional burden to wildfires in the US from 1980-2014. These are categorized by "fire weather zone" as defined by the National Weather Service.

Nationally, the burden of wildfires is disproportionally heavily distributed in the southern and western regions. The Geographic Area Coordinating Group (GACG) divides the United States and Alaska into 11 geographic areas for the purpose of emergency incident management. One particular area of focus is wildland fires. A national assessment of wildfire risk in the United States based on GACG identified regions (with the slight modification of combining Southern and Northern California, and the West and East

Basin); indicate that California (50.22% risk) and the Southern Area (15.53% risk) are the geographic areas with the highest wildfire risk. The western areas of the nation are experiencing an expansion of human development into and beyond what is called the wildland-urban interface (WUI). When wildfires inevitably occur in these fire-prone areas, often communities are threatened due to their proximity to fire-prone forest. The south is one of the fastest growing regions with 88 million acres classified as WUI. The south consistently has the highest number of wildfires per year. More than 50, 000 communities are estimated to be at high to very high risk of wildfire damage. These statistics are greatly attributable to the South's year-round fire season.

Wildfires' Risk to Human Health

The most noticeable adverse effect of wildfires is the destruction of property. However, the release of hazardous chemicals from the burning of wildland fuels also significantly impacts health in humans.

Wildfire smoke is composed primarily of carbon dioxide and water vapor. Other common smoke components present in lower concentrations are carbon monoxide, formaldehyde, acrolein, polyaromatic hydrocarbons, and benzene. Small particulates suspended in air which come in solid form or in liquid droplets are also present in smoke. 80 -90% of wildfire smoke, by mass, is within the fine particle size class of 2.5 micrometers in diameter or smaller.

Despite carbon dioxide's high concentration in smoke, it poses a low health risk due to its low toxicity. Rather, carbon monoxide and fine particulate matter, particularly 2.5 μm in diameter and smaller, have been identified as the major health threats. Other chemicals are considered to be significant hazards but are found in concentrations that are too low to cause detectable health effects.

The degree of wildfire smoke exposure to an individual is dependent on the length, severity, duration, and proximity of the fire. People are exposed directly to smoke via the respiratory tract though inhalation of air pollutants. Indirectly, communities are exposed to wildfire debris that can contaminate soil and water supplies.

Firefighters are at the greatest risk for acute and chronic health effects resulting from wildfire smoke exposure. Due to firefighters' occupational duties, they are frequently exposed to hazardous chemicals at a close proximity for longer periods of time. A case study on the exposure of wildfire smoke among wildland firefighters shows that firefighters are exposed to significant levels of carbon monoxide and respiratory irritants above OSHA-permissible exposure limits (PEL) and ACGIH threshold limit values (TLV). 5–10% are overexposed. The study obtained exposure concentrations for one wildland firefighter over a 10-hour shift spent holding down a fireline. The firefighter was exposed to a wide range of carbon monoxide and respiratory irritant (combination of particulate matter 3.5 μm and smaller, acrolein, and formaldehype) levels. Carbon monoxide levels reached up to 160ppm and the TLV irritant index value reached a high of 10. In contrast, the OSHA PEL for carbon monoxide is 30ppm and for the TLV respiratory irritant index, the calculated threshold limit value is 1; any value above 1 exceeds exposure limits.

Residents in communities surrounding wildfires are exposed to lower concentrations of chemicals, but they are at a greater risk for indirect exposure through water or soil contamination. Exposure to residents is greatly dependent on individual susceptibility. Vulnerable persons such as children (ages 0–4), the elderly (ages 65 and older), smokers, and pregnant women are at an increased risk due to their already compromised body systems, even when the exposures are present at low chemical concentrations and for relatively short exposure periods.

The U.S. Environmental Protection Agency (EPA) developed the Air Quality Index (AQI), a public resource that provides national air quality standard concentrations for common air pollutants. The public can use this index as a tool to determine their exposure to hazardous air pollutants based on visibility range.

Additionally, there is evidence of an increase in material stress, as documented by researchers M.H. O'Donnell and A.M. Behie, thus affecting birth outcomes. In Australia, studies show that male infants born with drastically higher average birth weights were born in mostly severely fire-affected areas. This is attributed to the fact that maternal signals directly affect fetal growth patterns.

Health Effects

Inhalation of smoke from a wildfire can be a health hazard. Wildfire smoke is primarily composed of carbon dioxide, water vapor, particulate matter, organic chemicals, nitrogen oxides and other compounds. The principle health concern is the inhalation of particulate matter and carbon monoxide.

Particulate matter (PM) is a type of air pollution made up of particles of dust and liquid droplets. They are characterized into two categories based on the diameter of the particle. Coarse particles are between 2.5 micrometers and 10 micrometers and fine particles measure 2.5 micrometers and less. Both sizes can be inhaled. Coarse particles are filtered by the upper airways and can cause eye and sinus irritation as well as sore throat and coughing. The fine particles are more problematic because, when inhaled, they can be deposited deep into the lungs, where they are absorbed into the bloodstream. This is particularly hazardous to the very young, elderly and those with chronic conditions such as asthma, chronic obstructive pulmonary disease (COPD), cystic fibrosis and cardiovascular conditions. The illnesses most commonly with exposure to fine particle from wildfire smoke are bronchitis, exacerbation of asthma or COPD, and pneumonia. Symptoms of these complications include wheezing and shortness of breath and cardiovascular symptoms include chest pain, rapid heart rate and fatigue.

Carbon monoxide (CO) is a colorless, odorless gas that can be found at the highest concentration at close proximity to a smoldering fire. For this reason, carbon monoxide inhalation is a serious threat to the health of wildfire firefighters. CO in smoke can be inhaled into the lungs where it is absorbed into the bloodstream and reduces oxygen delivery to the body's vital organs. At high concentrations, it can cause headache, weakness, dizziness, confusion, nausea, disorientation, visual impairment, coma and even death. However, even at lower concentrations, such as those found at wildfires, individuals with cardiovascular disease may experience chest pain and cardiac arrhythmia. A recent study tracking the number and cause of wildfire firefighter deaths from 1990–2006 found that 21.9% of the deaths occurred from heart attacks.

Another important and somewhat less obvious health effect of wildfires is psychiatric diseases and disorders. Both adults and children from countries ranging from the United States and Canada to Greece and Australia who were directly and indirectly affected by wildfires were found by researchers to demonstrate several different mental conditions linked to their experience with the wildfires. These include post-traumatic stress disorder (PTSD), depression, anxiety, and phobias.

In a new twist to wildfire health effects, former uranium mining sites were burned over in the summer of 2012 near North Fork, Idaho. This prompted concern from area residents and Idaho State Department of Environmental Quality officials over the potential spread of radiation in the resultant smoke, since those sites had never been completely cleaned up from radioactive remains.

Epidemiology

The EPA has defined acceptable concentrations of particulate matter in the air, through the National Ambient Air Quality Standards and monitoring of ambient air quality has been mandated. Due to these monitoring programs and the incidence of several large wildfires near populated areas, epidemiological studies have been conducted and demonstrate an association between human health effects and an increase in fine particulate matter due to wildfire smoke.

An increase in PM emitted from the Hayman fire in Colorado in June 2002, was associated with an increase in respiratory symptoms in patients with COPD. Looking at the wildfires in Southern California in October 2003 in a similar manner, investigators have shown an increase in hospital admissions due to asthma during peak concentrations of PM. Children participating in the Children's Health Study were also found to have an increase in eye and respiratory symptoms, medication use and physician visits. Recently, it was demonstrated that mothers who were pregnant during the fires gave birth to babies with a slightly reduced average birth weight compared to those who were not exposed to wildfire during birth. Suggesting that pregnant women may also be at greater risk to adverse effects from wildfire. Worldwide it is estimated that 339,000 people die due to the effects of wildfire smoke each year.

Invasive Species

Beavers from North America constitute an invasive species in Tierra del Fuego, where they have a substantial impact on landscape and local ecology through their dams.

Kudzu, a Japanese vine species invasive in the southeast United States, growing in Atlanta, Georgia

An invasive species is a plant, fungus, or animal species that is not native to a specific location (an introduced species), and which has a tendency to spread to a degree believed to cause damage to the environment, human economy or human health.

One study pointed out widely divergent perceptions of the criteria for invasive species among researchers (p. 135) and concerns with the subjectivity of the term "invasive" (p. 136). Some of the alternate usages of the term are below:

- The term as most often used applies to introduced species (also called "non-indigenous" or "non-native") that adversely affect the habitats and bioregions they invade economically, environmentally, or ecologically. Such invasive species may be either plants or animals and may disrupt by dominating a region, wilderness areas, particular habitats, or wildland-urban interface land from loss of natural controls (such as predators or herbivores). This includes non-native invasive plant species labeled as exotic pest plants and invasive exotics growing in native plant communities. It has been used in this sense by government organizations as well as conservation groups such as the International Union for

Conservation of Nature (IUCN) and the California Native Plant Society. The European Union defines "Invasive Alien Species" as those that are, firstly, outside their natural distribution area, and secondly, threaten biological diversity. It is also used by land managers, botanists, researchers, horticulturalists, conservationists, and the public for noxious weeds. The kudzu vine (*Pueraria lobata*), Andean Pampas grass (*Cortaderia jubata*), and yellow starthistle (*Centaurea solstitialis*) are examples.

- An alternate usage broadens the term to include indigenous or "native" species along with *non-native* species, that have colonized natural areas. Deer are an example, considered to be overpopulating their native zones and adja-cent suburban gardens, by some in the Northeastern and Pacific Coast regions of the United States.
- Sometimes the term is used to describe a non-native or introduced species that has become widespread. However, not every introduced species has adverse effects on the environment. A nonadverse example is the common goldfish (*Carassius auratus*), which is found throughout the United States, but rarely achieves high densities.

Causes

Scientists include species- and ecosystem factors among the mechanisms that when combined, establish invasiveness in a newly introduced species.

Species-based Mechanisms

While all species compete to survive, invasive species appear to have specific traits or specific combinations of traits that allow them to outcompete native species. In some cases, the competition is about rates of growth and reproduction. In other cases, species interact with each other more directly.

Researchers disagree about the usefulness of traits as invasiveness markers. One study found that of a list of invasive and noninvasive species, 86% of the invasive species could be identified from the traits alone. Another study found invasive species tended to have only a small subset of the presumed traits and that many similar traits were found in noninvasive species, requiring other explanations. Common invasive species traits include the following:

- Fast growth
- Rapid reproduction
- High dispersal ability
- Phenotypic plasticity (the ability to alter growth form to suit current conditions)

- Tolerance of a wide range of environmental conditions (Ecological competence)
- Ability to live off of a wide range of food types (generalist)
- Association with humans
- Prior successful invasions

Typically, an introduced species must survive at low population densities before it becomes invasive in a new location. At low population densities, it can be difficult for the introduced species to reproduce and maintain itself in a new location, so a species might reach a location multiple times before it becomes established. Repeated patterns of human movement, such as ships sailing to and from ports or cars driving up and down highways offer repeated opportunities for establishment (also known as a high propagule pressure).

An introduced species might become invasive if it can outcompete native species for resources such as nutrients, light, physical space, water, or food. If these species evolved under great competition or predation, then the new environment may host fewer able competitors, allowing the invader to proliferate quickly. Ecosystems in which are being used to their fullest capacity by native species can be modeled as zero-sum systems in which any gain for the invader is a loss for the native. However, such unilateral competitive superiority (and extinction of native species with increased populations of the invader) is not the rule. Invasive species often coexist with native species for an extended time, and gradually, the superior competitive ability of an invasive species becomes apparent as its population grows larger and denser and it adapts to its new location.

Lantana growing in abandoned citrus plantation; Moshav Sdei Hemed, Israel

An invasive species might be able to use resources that were previously unavailable to native species, such as deep water sources accessed by a long taproot, or an ability to live on previously uninhabited soil types. For example, barbed goatgrass (*Aegilops triuncialis*) was introduced to California on serpentine soils, which have low water-retention, low nutrient levels, a high magnesium/calcium ratio, and possible heavy metal toxicity. Plant populations on these soils tend to show low density, but goatgrass can form dense stands on these soils and crowd out native species that have adapted poorly to serpentine soils.

Invasive species might alters its environment by releasing chemical compounds, modifying abiotic factors, or affecting the behaviour of herbivores, creating a positive or negative impact on other species. Some species, like *Kalanchoe daigremontana*, produce allelopathic compounds, that might have an inhibitory effect on competing species. Other species like *Stapelia gigantea* facilitates the recruitment of seedlings of other species in arid environments by providing appropriate microclimatic conditions and preventing herbivory in early stages of development.

Another examples are *Centaurea solstitialis* (yellow starthistle) and *Centaurea diffusa* (diffuse knapweed). These Eastern European noxious weeds have spread through the western and West Coast states. Experiments show that 8-hydroxyquinoline, a chemical produced at the root of *C. diffusa*, has a negative effect only on plants that have not co-evolved with it. Such co-evolved native plants have also evolved defenses. *C. diffusa* and *C. solstitialis* do not appear in their native habitats to be overwhelmingly successful competitors. Success or lack of success in one habitat does not necessarily imply success in others. Conversely, examining habitats in which a species is less successful can reveal novel weapons to defeat invasiveness.

Changes in fire regimens are another form of facilitation. *Bromus tectorum*, originally from Eurasia, is highly fire-adapted. It not only spreads rapidly after burning but also increases the frequency and intensity (heat) of fires by providing large amounts of dry detritus during the fire season in western North America. In areas where it is widespread, it has altered the local fire regimen so much that native plants cannot survive the frequent fires, allowing *B. tectorum* to further extend and maintain dominance in its introduced range.

Facilitation also occurs where one species physically modifies a habitat in ways that are advantageous to other species. For example, zebra mussels increase habitat complexity on lake floors, providing crevices in which invertebrates live. This increase in complexity, together with the nutrition provided by the waste products of mussel filter-feeding, increases the density and diversity of benthic invertebrate communities.

Ecosystem-based Mechanisms

In ecosystems, the amount of available resources and the extent to which those resources are used by organisms determines the effects of additional species on the ecosystem. In stable ecosystems, equilibrium exists in the use of available resources. These mechanisms describe a situation in which the ecosystem has suffered a disturbance, which changes the fundamental nature of the ecosystem.

When changes such as a forest fire occur, normal succession favors native grasses and forbs. An introduced species that can spread faster than natives can use resources that would have been available to native species, squeezing them out. Nitrogen and phosphorus are often the limiting factors in these situations.

Every species occupies a *niche* in its native ecosystem; some species fill large and varied roles, while others are highly specialized. Some invading species fill niches that are not used by native species, and they also can create new niches. An example of this type can be found within the *Lampropholis delicata* species of skink.

Ecosystem changes can alter species' distributions. For example, edge effects describe what happens when part of an ecosystem is disturbed as when land is cleared for agriculture. The boundary between remaining undisturbed habitat and the newly cleared land itself forms a distinct habitat, creating new winners and losers and possibly hosting species that would not thrive outside the boundary habitat.

One interesting finding in studies of invasive species has shown that introduced populations have great potential for rapid adaptation and this is used to explain how so many introduced species are able to establish and become invasive in new environments. When bottlenecks and founder effects cause a great decrease in the population size, the individuals begin to show additive variance as opposed to epistatic variance. This conversion can actually lead to increased variance in the founding populations which then allows for rapid adaptive evolution. Following invasion events, selection may initially act on the capacity to disperse as well as physiological tolerance to the new stressors in the environment. Adaptation then proceeds to respond to the selective pressures of the new environment. These responses would most likely be due to temperature and climate change, or the presence of native species whether it be predator or prey. Adaptations include changes in morphology, physiology, phenology, and plasticity.

Rapid adaptive evolution in these species leads to offspring that have higher fitness and are better suited for their environment. Intraspecific phenotypic plasticity, pre-adaptation and post-introduction evolution are all major factors in adaptive evolution. Plasticity in populations allows room for changes to better suit the individual in its environment. This is key in adaptive evolution because the main goal is how to best be suited to the ecosystem that the species has been introduced. The ability to accomplish this as quickly as possible will lead to a population with a very high fitness. Pre-adaptations and evolution after the initial introduction also play a role in the success of the introduced species. If the species has adapted to a similar ecosystem or contains traits that happen to be well suited to the area that it is introduced, it is more likely to fare better in the new environment. This, in addition to evolution that takes place after introduction, all determine if the species will be able to become established in the new ecosystem and if it will reproduce and thrive.

Ecology

Traits of Invaded Ecosystems

In 1958, Charles S. Elton claimed that ecosystems with higher species diversity were less subject to invasive species because of fewer available niches. Other ecologists later pointed to highly diverse, but heavily invaded ecosystems and argued that ecosystems with high species diversity were more susceptible to invasion.

This debate hinged on the spatial scale at which invasion studies were performed, and the issue of how diversity affects susceptibility remained unresolved as of 2011. Small-scale studies tended to show a negative relationship between diversity and invasion, while large-scale studies tended to show the reverse. The latter result may be a side-effect of invasives' ability to capitalize on increased resource availability and weaker species interactions that are more common when larger samples are considered.

The brown tree snake (*Boiga irregularis*)

Invasion was more likely in ecosystems that were similar to the one in which the potential invader evolved. Island ecosystems may be more prone to invasion because their species faced few strong competitors and predators, or because their distance from colonizing species populations makes them more likely to have "open" niches. An example of this phenomenon was the decimation of native bird populations on Guam by the invasive brown tree snake. Conversely, invaded ecosystems may lack the natural competitors and predators that check invasives' growth in their native ecosystems.

Invaded ecosystems may have experienced disturbance, typically human-induced. Such a disturbance may give invasive species a chance to establish themselves with less competition from natives less able to adapt to a disturbed ecosystem.

Vectors

Non-native species have many *vectors*, including biogenic vectors, but most invasions are associated with human activity. Natural range extensions are common in many species, but the rate and magnitude of human-mediated extensions in these species tend to be much larger than natural extensions, and humans typically carry specimens greater distances than natural forces.

An early human vector occurred when prehistoric humans introduced the Pacific rat (*Rattus exulans*) to Polynesia.

Chinese mitten crab (*Eriocheir sinensis*)

Vectors include plants or seeds imported for horticulture. The pet trade moves animals across borders, where they can escape and become invasive. Organisms stow away on transport vehicles. Ballast water taken up at sea and released in port by transoceanic vessels is the largest vector for non-native aquatic species invasions. Around the world on the average day, more than 3,000 different species of aquatic life may be transported on these vessels. For example, freshwater zebra mussels, native to the Black, Caspian and Azov seas, probably reached the Great Lakes via ballast water from a transoceanic vessel. Although the zebra mussel invasion was first noted in 1988, and a mitigation plan was successfully implemented shortly thereafter, the plan had (and continued to have as of 2005) a serious flaw or loophole, whereby ships that are loaded with cargo when they reach the Seaway need not be tested, but all the same they transfer ballast 'puddles' between Seaway ports.

The arrival of invasive propagules to a new site is a function of the site's invasibility.

Species have also been introduced intentionally. For example, to feel more "at home," American colonists formed "Acclimation Societies" that repeatedly imported birds that were native to Europe to North America and other distant lands. In 2008, U.S. postal workers in Pennsylvania noticed noises coming from inside a box from Taiwan; the box contained more than two dozen live beetles. Agricultural Research Service entomologists identified them as rhinoceros beetle, hercules beetle, and king stag beetle. Because these species were not native to the U.S., they could have threatened native ecosystems. To prevent exotic species from becoming a problem in the U.S., special handling and permits are required when living materials are shipped from foreign countries. USDA programs such as Smuggling Interdiction and Trade Compliance (SITC) attempt to prevent exotic species outbreaks in America.

Economics plays a major role in exotic species introduction. High demand for the valuable Chinese mitten crab is one explanation for the possible intentional release of the species in foreign waters.

Impacts of Wildfire

Invasive species often exploit disturbances to an ecosystem (wildfires, roads, foot trails) to colonize an area. Large wildfires can sterilize soils, while adding a variety of nutrients. In the resulting free-for-all, formerly entrenched species lose their advantage, leaving more room for invasives. In such circumstances plants that can regenerate from their roots have an advantage. Non-natives with this ability can benefit from a low intensity fire burns that removes surface vegetation, leaving natives that rely on seeds for propagation to find their niches occupied when their seeds finally sprout.

Impact of Wildfire Suppression on Spreading

Wildfires often occur in remote areas, needing fire suppression crews to travel through pristine forest to reach the site. The crews can bring invasive seeds with them. If any of these stowaway seeds become established, a thriving colony of invasives can erupt in as few as six weeks, after which controlling the outbreak can need years of continued attention to prevent further spread. Also, disturbing the soil surface, such as cutting firebreaks, destroys native cover, exposes soil, and can accelerate invasions. In suburban and wildland-urban interface areas, the vegetation clearance and brush removal ordinances of municipalities for defensible space can result in excessive removal of native shrubs and perennials that exposes the soil to more light and less competition for invasive plant species.

Fire suppression vehicles are often major culprits in such outbreaks, as the vehicles are often driven on back roads often overgrown with invasive plant species. The undercarriage of the vehicle becomes a prime vessel of transport. In response, on large fires, washing stations "decontaminate" vehicles before engaging in suppression activities. Large wildfires attract firefighters from remote places, further increasing the potential for seed transport.

Effects

An American alligator attacking a Burmese python in Florida; the Burmese python is an invasive species which is posing a threat to many indigenous species, including the alligator

Ecological

Land clearing and human habitation put significant pressure on local species. Disturbed habitats are prone to invasions that can have adverse effects on local ecosystems, changing ecosystem functions. A species of wetland plant known as 'ae' ae in Hawaii (the indigenous *Bacopa monnieri*) is regarded as a pest species in artificially manipulated water bird refuges because it quickly covers shallow mudflats established for endangered Hawaiian stilt (*Himantopus mexicanus knudseni*), making these undesirable feeding areas for the birds.

Multiple successive introductions of different non-native species can have interactive effects; the introduction of a second non-native species can enable the first invasive species to flourish. Examples of this are the introductions of the amethyst gem clam (*Gemma gemma*) and the European green crab (*Carcinus maenas*). The gem clam was introduced into California's Bodega Harbor from the East Coast of the United States a century ago. It had been found in small quantities in the harbor but had never displaced the native clam species (*Nutricola* spp.). In the mid-1990s, the introduction of the European green crab, found to prey preferentially on the native clams, resulted in a decline of the native clams and an increase of the introduced clam populations.

In the Waterberg region of South Africa, cattle grazing over the past six centuries has allowed invasive scrub and small trees to displace much of the original grassland, resulting in a massive reduction in forage for native bovids and other grazers. Since the 1970s, large scale efforts have been underway to reduce invasive species; partial success has led to re-establishment of many species that had dwindled or left the region. Examples of these species are giraffe, blue wildebeest, impala, kudu and white rhino.

Invasive species can change the functions of ecosystems. For example, invasive plants can alter the fire regimen (cheatgrass, *Bromus tectorum*), nutrient cycling (smooth cordgrass *Spartina alterniflora*), and hydrology (*Tamarix*) in native ecosystems. Invasive species that are closely related to rare native species have the potential to hybridize with the native species. Harmful effects of hybridization have led to a decline and even extinction of native species. For example, hybridization with introduced cordgrass, *Spartina alterniflora*, threatens the existence of California cordgrass (*Spartina foliosa*) in San Francisco Bay. Invasive species cause competition for native species and because of this 400 of the 958 endangered species under the Endangered Species Act are at risk

Geomorphological

Primary geomorphological effects of invasive plants are bioconstruction and bioprotection. For example, Kudzu Pueraria montana, a vine native to Asia was widely introduced in the southeastern USA in the early 20th century to control soil erosion. While primary effects of invasive animals are bioturbation, bioerosion, and bioconstruction.

For example, invasion of Chinese mitten crab Eriocheir sinensis have resulted in higher bioturbation and bioerosion rates.

Economic

Benefits

Non-native species can have benefits. Asian oysters, for example, filter water pollutants better than native oysters. They also grow faster and withstand disease better than natives. Biologists are currently considering releasing this mollusk in the Chesapeake Bay to help restore oyster stocks and remove pollution. A recent study by the Johns Hopkins School of Public Health found the Asian oyster could significantly benefit the bay's deteriorating water quality. Additionally, some species have invaded an area so long ago that they have found their own beneficial niche in the environment. For example, *L. leucozonium*, shown by population genetic analysis to be an invasive species in North America, has become an important pollinator of caneberry as well as cucurbit, apple trees, and blueberry bushes.

Costs

Economic costs from invasive species can be separated into direct costs through production loss in agriculture and forestry, and management costs. Estimated damage and control cost of invasive species in the U.S. alone amount to more than $138 billion annually. Economic losses can also occur through loss of recreational and tourism revenues. When economic costs of invasions are calculated as production loss and management costs, they are low because they do not consider environmental damage; if monetary values were assigned to the extinction of species, loss in biodiversity, and loss of ecosystem services, costs from impacts of invasive species would drastically increase. The following examples from different sectors of the economy demonstrate the impact of biological invasions.

Economic Opportunities

Some invasions offer potential commercial benefits. For instance, silver carp and common carp can be harvested for human food and exported to markets already familiar with the product, or processed into pet foods, or mink feed. Vegetative invasives such as water hyacinth can be turned into fuel by methane digesters .

Invasivorism

Invasive species are flora and fauna whose introduction into a habitat disrupts the native eco-system. In response, Invasivorism is a movement that explores the idea of eating invasive species in order to control, reduce, or eliminate their populations. Chefs from around the world have begun seeking out and using invasive species as alternative ingredients. Miya's of New Haven, Connecticut created the first invasive species menu in the world. Skeptics point out that once a foreign species has entrenched itself in a new place—such as the Indo-Pacific lionfish that has now virtually taken over the waters of the Western Atlantic, Caribbean and Gulf of Mexico—eradication is almost impossible. Critics argue that encouraging consumption might have the unintended effect of spreading harmful species even more widely.

A dish that features whole fried invasive lionfish at Fish Fish of Miami, Florida

Proponents of invasivorism argue that humans have the ability to eat away any species that it has an appetite for, pointing to the many animals which humans have been able to hunt to extinction - such as the Dodo bird, the Caribbean monk seal, and the Passenger pigeon. Proponents of invasivorism also point to the success that Jamaica has had in significantly decreasing the population of lionfish by encouraging the consumption of the fish.

Plant Industry

Weeds reduce yield in agriculture, though they may provide essential nutrients. Some deep-rooted weeds can "mine" nutrients from the subsoil and deposit them on the topsoil, while others provide habitat for beneficial insects or provide foods for pest species. Many weed species are accidental introductions that accompany seeds and imported plant material. Many introduced weeds in pastures compete with native forage plants, threaten young cattle (e.g., leafy spurge, *Euphorbia esula*) or are unpalatable because of thorns and spines (e.g., yellow starthistle). Forage loss from invasive weeds on pastures amounts to nearly US$1 billion in the U.S. alone. A decline in pollinator services and loss of fruit production has been caused by honey bees infected by the invasive varroa mite. Introduced rats (*Rattus rattus* and *R. norvegicus*) have become serious pests on farms, destroying stored grains.

Invasive plant pathogens and insect vectors for plant diseases can also suppress agricultural yields and nursery stock. Citrus greening is a bacterial disease vectored by the invasive Asian citrus psyllid (ACP). Because of the impacts of this disease on citrus crops, citrus is under quarantine and highly regulated in areas where ACP has been found.

Aquaculture

Aquaculture is a very common vector of species introductions – mainly of species with economic potential (e.g., Oreochromis niloticus)

Forestry

Poster asking campers to not move firewood around, avoiding the spread of invasive species.

The unintentional introduction of forest pest species and plant pathogens can change forest ecology and damage the timber industry. Overall, forest ecosystems in the U.S. are widely invaded by exotic pests, plants, and pathogens.

The Asian long-horned beetle (*Anoplophora glabripennis*) was first introduced into the U.S. in 1996, and was expected to infect and damage millions of acres of hardwood trees. As of 2005 thirty million dollars had been spent in attempts to eradicate this pest and protect millions of trees in the affected regions. The woolly adelgid has inflicted damage on old-growth spruce, fir and hemlock forests and damages the Christmas tree industry. And the chestnut blight fungus (*Cryphonectria parasitica*) and Dutch elm disease (*Ophiostoma novo-ulmi*) are two plant pathogens with serious impacts on these two species, and forest health. Garlic mustard, *Alliaria petiolata,* is one of the most problematic invasive plant species in eastern North American forests. The characteristics of garlic mustard are slightly different from those of the surrounding native plants, which results in a highly successful species that is altering the composition and function of the native communities it invades. When garlic mustard invades the understory of a forest, it affects the growth rate of tree seedlings, which is likely to alter forest regeneration of impact forest composition in the future.

Tourism and Recreation

Invasive species can impact outdoor recreation, such as fishing, hunting, hiking, wildlife viewing, and water-based activities. They can damage a wide array of environmental services that are important to recreation, including, but not limited to, water quality and quantity, plant and animal diversity, and species abundance. Eiswerth states, "very little research has been performed to estimate the corresponding economic losses at spatial scales such as regions, states, and watersheds." Eurasian watermilfoil (*Myriophyllum spicatum*) in parts of the US, fill lakes with plants complicating fishing and boating. The very loud call of the introduced common coqui depresses real estate values in affected neighborhoods of Hawaii.

Health

Encroachment of humans into previously remote ecosystems has exposed exotic diseases such as HIV to the wider population. Introduced birds (e.g. pigeons), rodents and insects (e.g. mosquito, flea, louse and tsetse fly pests) can serve as vectors and reservoirs of human afflictions. The introduced Chinese mitten crabs are carriers of Asian lung fluke. Throughout recorded history, epidemics of human diseases, such as malaria, yellow fever, typhus, and bubonic plague, spread via these vectors. A recent example of an introduced disease is the spread of the West Nile virus, which killed humans, birds, mammals, and reptiles. Waterborne disease agents, such as cholera bacteria (*Vibrio cholerae*), and causative agents of harmful algal blooms are often transported via ballast water. Invasive species and accompanying control efforts can have long term public health implications. For instance, pesticides applied to treat a particular pest species could pollute soil and surface water.

Biodiversity

Biotic invasion is considered one of the five top drivers for global biodiversity loss and is increasing because of tourism and globalization. This may be particularly true in inadequately regulated fresh water systems, though quarantines and ballast water rules have improved the situation.

Invasive species may drive local native species to extinction via competitive exclusion, niche displacement, or hybridisation with related native species. Therefore, besides their economic ramifications, alien invasions may result in extensive changes in the structure, composition and global distribution of the biota of sites of introduction, leading ultimately to the homogenisation of the world's fauna and flora and the loss of biodiversity. Nevertheless, it is difficult to unequivocally attribute extinctions to a species invasion, and the few scientific studies that have done so have been with animal taxa. Concern over the impacts of invasive species on biodiversity must therefore consider the actual evidence (either ecological or economic), in relation to the potential risk.

Genetic Pollution

Native species can be threatened with extinction through the process of *genetic pollution*. Genetic pollution is unintentional hybridization and introgression, which leads to homogenization or replacement of local genotypes as a result of either a numerical or fitness advantage of the introduced species. Genetic pollution can operate either through introduction or through habitat modification, bringing previously isolated species into contact. Hybrids resulting from rare species that interbreed with abundant species can swamp the rarer species' gene pool. This is not always apparent from morphological observations alone. Some degree of gene flow is normal, and preserves constellations of genes and genotypes. An example of this is the interbreeding of migrating coyotes with the red wolf, in areas of eastern North Carolina where the red wolf was reintroduced.

Study

Stage	Characteristic
0	Propagules residing in a donor region
I	Traveling
II	Introduced
III	Localized and numerically rare
IVa	Widespread but rare
IVb	Localized but dominant
V	Widespread and dominant

While the study of invasive species can be done within many subfields of biology, the majority of research on invasive organisms has been within the field of ecology and geography where the issue of biological invasions is especially important. Much of the study of invasive species has been influenced by Charles Elton's 1958 book *The Ecology of Invasion by Animals and Plants* which drew upon the limited amount of research done within disparate fields to create a generalized picture of biological invasions. Studies on invasive species remained sparse until the 1990s when research in the field experienced a large amount of growth which continues to this day. This research, which has largely consisted of field observational studies, has disproportionately been concerned with terrestrial plants. The rapid growth of the field has driven a need to standardize the language used to describe invasive species and events. Despite this, little standard terminology exists within the study of invasive species which itself lacks any official designation but is commonly referred to as "Invasion ecology" or more generally "Invasion biology". This lack of standard terminology is a significant problem, and has largely arisen due to the interdisciplinary nature of the field which borrows terms from numerous disciplines such as agriculture, zoology, and pathology, as well as due to studies on invasive species being commonly performed in isolation of one another.

In an attempt to avoid the ambiguous, subjective, and pejorative vocabulary that so often accompanies discussion of invasive species even in scientific papers, Colautti and MacIsaac proposed a new nomenclature system based on biogeography rather than on taxa.

By discarding taxonomy, human health, and economic factors, this model focused only on ecological factors. The model evaluated individual populations rather than entire species. It classified each population based on its success in that environment. This model applied equally to indigenous and to introduced species, and did not automatically categorize successful introductions as harmful.

References

- Rudel, T.K. 2005 "Tropical Forests: Regional Paths of Destruction and Regeneration in the Late 20th Century" Columbia University Press ISBN 0-231-13195-X
- Leakey, Richard and Roger Lewin, 1996, The Sixth Extinction : Patterns of Life and the Future of Humankind, Anchor, ISBN 0-385-46809-1
- Taylor, Leslie (2004). The Healing Power of Rainforest Herbs: A Guide to Understanding and Using Herbal Medicinals. Square One. ISBN 9780757001444.
- Norman F. Cantor (9 June 1994). The civilization of the Middle Ages: a completely revised and expanded edition of Medieval history, the life and death of a civilization. HarperCollins. p. 564. ISBN 978-0-06-092553-6. Retrieved 4 December 2011.
- Ron Nielsen, The Little Green Handbook: Seven Trends Shaping the Future of Our Planet, Picador, New York (2006) ISBN 978-0-312-42581-4
- John F. Mongillo; Linda Zierdt-Warshaw (2000). Linda Zierdt-Warshaw, ed. Encyclopedia of environmental science. University of Rochester Press. p. 104. ISBN 978-1-57356-147-1.
- Daniel B. Botkin (2001). No man's garden: Thoreau and a new vision for civilization and nature. Island Press. pp. 246–247. ISBN 978-1-55963-465-6. Retrieved 4 December 2011.
- M. Patricia Marchak (18 September 1995). Logging the globe. McGill-Queen's Press – MQUP. pp. 157–. ISBN 978-0-7735-1346-4. Retrieved 4 December 2011.
- Pyne, Stephen J.; Andrews, Patricia L.; Laven, Richard D. (1996). Introduction to wildland fire (2nd ed.). John Wiley and Sons. p. 65. ISBN 978-0-471-54913-0. Retrieved 26 January 2010.
- Elton, C.S. (2000) [1958]. The Ecology of Invasions by Animals and Plants. Foreword by Daniel Simberloff. Chicago: University of Chicago Press. p. 196. ISBN 0-226-20638-6.
- Bill Gabbert (November 9, 2015). "Was the 2014 wildfire season in California affected by climate change?". Wildfire Today. Retrieved May 17, 2016.

6

Analytical Aspects and Developments of Forest Conservation

Restoration ecology, conservation movement and conservation biology are the vital aspects of forest conservation. Restoration ecology is the practice of restoring destroyed ecosystems whereas the conservation movement is a social movement that pursues to safeguard the environment. This section elucidates the aspects and developments of forest conservation.

Restoration Ecology

Restoration ecology emerged as a separate field in ecology in the 1980s. It is the scientific study supporting the practice of ecological restoration, which is the practice of renewing and restoring degraded, damaged, or destroyed ecosystems and habitats in the environment by active human intervention and action. The term "restoration ecology" is therefore commonly used for the academic study of the process, whereas the term "ecological restoration" is commonly used for the actual project or process by restoration practitioners.

Rehabilitation of a portion of Johnson Creek, to restore bioswale and flood control functions of the land which had long been converted to pasture for cow grazing. The horizontal logs can float, but are anchored by the posts. Just-planted trees will eventually stabilize the soil. The fallen trees with roots jutting into the stream are intended to enhance wildlife habitat. The meandering of the stream is enhanced here by a factor of about three times, perhaps to its original course.

Definition

The Society for Ecological Restoration defines *"ecological restoration"* as an "intentional activity that initiates or accelerates the recovery of an ecosystem with respect to its health, integrity and sustainability". The practice of ecological restoration includes wide scope of projects such as erosion control, reforestation, usage of genetically local native species, removal of non-native species and weeds, revegetation of disturbed areas, daylighting streams, reintroduction of native species, as well as habitat and range improvement for targeted species.

E. O. Wilson, a biologist states that: "Here is the means to end the great extinction spasm. The next century will, I believe, be the era of restoration in ecology."

Restoration ecology is the scientific study of ecological restoration.

History

Land managers, laypeople, and stewards have been practicing ecological restoration or ecological management for many hundreds, if not thousands of years, yet the scientific field of "restoration ecology" was not first formally identified and coined until the late 1980s, by John Aber and William Jordan when they were at the University of Wisconsin-Madison. They held the first international meetings on this topic in Madison during which attendees visited the University of Wisconsin's Arboretum—the oldest restoration ecology project made famous by Professor Aldo Leopold. The study of restoration ecology has only become a robust and independent scientific discipline over the last two decades, and the commercial applications of ecological restoration have tremendously increased in recent years.

Restoration Needs

There is consensus in the scientific community that the current environmental degradation and destruction of many of the Earth's biota is considerable and is taking place on a "catastrophically short timescale". Estimates of the current extinction rate is 1000 to 10,000 times more than the normal rate. For many people biological diversity, (biodiversity) has an intrinsic value that humans have a responsibility towards other living things, and an obligation to future generations.

On a more anthropocentric level, natural ecosystems provide human society with food, fuel and timber. Fundamentally, ecosystem services involve the purification of air and water, detoxification and decomposition of wastes, regulation of climate, regeneration of soil fertility and pollination of crops. Such processes have been estimated to be worth trillions of dollars annually.

Habitat loss is the leading cause of both species extinctions and ecosystem service decline. The two ways to reverse this trend of habitat loss are conservation of currently viable habitat and restoration of degraded habitats.

Conservation Biology and Restoration Ecology

Restoration ecology may be viewed as a sub-discipline of conservation biology, the scientific study of how to protect and restore biodiversity, and restoration a part of the resulting conservation movement.

Conservation vs. Restoration

The fundamental difference between restoration and other conservation efforts is analogous to the difference between disease prevention and treatment. Conservation attempts to maintain and protect existing habitat and biodiversity, whereas restoration attempts to reverse existing environmental degradation and population declines. Targeted human intervention is used to promote habitat, biodiversity recovery and associated gains.

The possibility of restoration, however, does not provide an excuse for converting extremely valuable "pristine" habitat into other uses: as in medicine, it better to prevent than to treat. "Treatment" is generally less effective and more expensive than prevention, and "treatment" cannot always restore the condition before the "injury": some habitat and biodiversity losses are permanent.

Focuses

Though restoration ecologists and other conservation biologists generally agree that habitat is the most important locus of biodiversity protection, the disciplines themselves have different focuses. Conservation biology as an academic discipline is rooted in population biology. Because of that, it is generally organized at the genetic level, looking at specific species populations (i.e. endangered species). Restoration ecology is organized at the community level, looking at specific ecosystems.

Because it is organized by species, conservation biology often emphasizes vertebrate animals because of their salience and popularity, whereas restoration ecology emphasizes plants because of their importance within ecosystems: ecosystem restoration is botanically based. Since soils define the foundation of any functional terrestrial system, restoration ecology's ecosystem-level focus also results in greater emphasis on the role of soil's physical and microbial processes.

Modes of Inquiry

Conservation biology's focus on rare or endangered species limit the number of manipulative studies that can be performed. As a consequence, conservation studies tend to be descriptive, comparative and unreplicable. However, the highly manipulative nature of restoration ecology allows the researcher to test the hypotheses vigorously. Restorative activity often reflects an experimental test of what limits populations.

Theoretical Foundations

Restoration ecology draws on a wide range of ecological concepts.

Disturbance

Disturbance is a change of environmental conditions, which interferes with the functioning of a biological system. Disturbance, at a variety of spatial and temporal scales is a natural, and even essential, component of many communities.

Humans have had limited "natural" impacts on ecosystems for as long as humans have existed, however, the severity and scope of our influences has accelerated in the last few centuries. Understanding and minimizing the differences between modern anthropogenic and "natural" disturbances is crucial to restoration ecology. For example, new forestry techniques that better imitate historical disturbances are now being implemented.

In addition, restoring a fully sustainable ecosystem often involves studying and attempting to restore a natural disturbance regime (e.g., fire ecology).

Succession

Ecological succession is the process by which the component species of a community changes over time. Following a disturbance, an ecosystem generally progresses from a simple level of organization (i.e. few dominant species) to a more complex community (i.e. many interdependent species) over few generations. Depending on the severity of the disturbance, restoration often consists of initiating, assisting or accelerating ecological successional processes.

In many ecosystems, communities tend to recover following mild to moderate natural and anthropogenic disturbances. Restoration in these systems involves hastening natural successional trajectories. However, a system that has experienced a more severe disturbance (i.e. physical or chemical alteration of the environment) may require intensive restorative efforts to recreate environmental conditions that favor natural successional processes. This ability to recover is called resilience.

Fragmentation

Habitat fragmentation is the emergence of spatial discontinuities in a biological system. Through land use changes (e.g. agriculture) and "natural" disturbance, ecosystems are broken up into smaller parts. Small fragments of habitat can support only small populations and small populations are more vulnerable to extinction. Furthermore, fragmenting ecosystems decreases interior habitat. Habitat along the edge of a fragment has a different range of environmental conditions and therefore supports different species than the interior. Fragmentation effectively reduces interior habitat and may lead to the extinction of those species which require interior habitat. Restorative projects can increase the effective size of a habitat by simply adding area or by planting habitat corridors that link and fill in the gap between two isolated fragments. Reversing the effects of fragmentation and increasing habitat connectivity are the central goals of restoration ecology.

Ecosystem Function

Ecosystem function describes the foundational processes of natural systems, including nutrient cycles and energy fluxes. These processes are the most basic and essential components of ecosystems. An understanding of the full complexity and intricacies of these cycles is necessary to address any ecological processes that may be degraded. A functional ecosystem, that is completely self-perpetuating (no management required), is the ultimate goal of restorative efforts. Since, these ecosystem functions are emergent properties of the system as a whole, monitoring and management are crucial for the long-term stability of an ecosystem.

Evolving Concepts

Restoration ecology, because of its highly physical nature, is an ideal testing ground for an emerging community's ecological principles (Bradshaw 1987). Likewise, there are emerging concepts of inventing new and successful restoration technologies, performance standards, time frames, local genetics, and society's relationship to restoration ecology, and new ethical and religious possibilities, as future topics of discussion and debate.

Assembly

Community assembly "is a framework that can unify virtually all of (community) ecology under a single conceptual umbrella". Community assembly theory attempts to explain the existence of environmentally similar sites with differing assemblages of species. It assumes that species have similar niche requirements, so that community formation is a product of random fluctuations from a common species pool. Essentially, if all species are fairly ecologically equivalent then random variation in colonization, migration and extinction rates between species, drive differences in species composition between sites with comparable environmental conditions.

Stable States

Alternative stable states are discrete species compositional possibilities that may exist within a community. According to assembly theory, differences in species colonization, interspecific interactions and community establishment may result in distinct community species equilibria. A community has numerous possible compositional equilibria that are dependent on the initial assembly. That is, random fluctuations lead to a particular initial community assembly, which affects successional trajectories and the eventual species composition equilibrium.

Multiple stable states is a specific theoretical concept, where all species have equal access to a community (i.e., equal dispersal potential) and differences between communities arise simply because of the timing of each species' colonization.

These concepts are central to restoration ecology; restoring a community not only involves manipulating the timing and structure of the initial species composition, but also working towards a single desired stable state. In fact, a degraded ecosystem may be viewed as an alternative stable state under the altered environmental conditions.

Ontogeny

The ecology of ontogeny is the study of how ecological relationships change over the lifetime of an individual. Organisms require different environmental conditions during different stages of their life-cycle. For immobile organisms (e.g. plants), the conditions necessary for germination and establishment may be different from those of the adult stage. As an ecosystem is altered by anthropogenic processes the range of environmental variables may also be altered. A degraded ecosystem may not include the environmental conditions necessary for a particular stage of an organism's development. If a self-sustaining, functional ecosystem must contain environmental conditions for the perpetual reproduction of its species, restorative efforts must address the needs of organisms throughout their development.

Application of Theory

Restoration is defined as the application of ecological theory to ecological restoration. However, for many reasons, this can be a challenging prospect. Here are a few examples of theory informing practice.

Soil Heterogeneity Effects on Community Heterogeneity

Spatial heterogeneity of resources can influence plant community composition, diversity and assembly trajectory. Baer et al. (2005) manipulated soil resource heterogeneity in a tallgrass prairie restoration project. They found increasing resource heterogeneity, which on its own was insufficient to insure species diversity in situations where one species may dominate across the range of resource levels. Their findings were consistent with the theory regarding the role of ecological filters on community assembly. The establishment of a single species, best adapted to the physical and biological conditions can play an inordinately important role in determining the community structure.

Invasion, Competitive Dominance and Resource Use

"The dynamics of invasive species may depend on their abilities to compete for resources and exploit disturbances relative to the abilities of native species". Seabloom et al. (2003) tested this concept and its implications in a California grassland restoration context. They found that the native grass species were able to successfully compete with invasive exotics, therefore, the possibility exists of restoring an original native grassland ecosystem.

Successional Trajectories

Progress along a desired successional pathway may be difficult if multiple stable states exist. Looking over 40 years of wetland restoration data Klotzi and Gootjans (2001) argue that unexpected and undesired vegetation assemblies "may indicate that environmental conditions are not suitable for target communities". Succession may move in unpredicted directions, but constricting environmental conditions within a narrow range may rein in the possible successional trajectories and increase the likelihood of a desired outcome.

The Natural Capital Committee's Recommendation for a 25-year Plan

The UK Natural Capital Committee.org/ Natural Capital Committee (NCC) made a recommendation in its second State of Natural Capital report (link) published in March 2014 that in order to meet the Government's goal of being the first generation to leave the environment in a better state than it was inherited, a long-term 25-year plan was needed to maintain and improve England's natural capital. The UK Government has not yet responded to this recommendation.

The Secretary of State for the UK's Department for Environment, Food and Rural Affairs, Owen Paterson, described his ambition for the natural environment and how the work of the Committee fits into this at an NCC event in November 2012: "I do not, however, just want to maintain our natural assets; I want to improve them. I want us to derive the greatest possible benefit from them, while ensuring that they are available for generations to come. This is what the NCC's innovative work is geared towards".

Ecosystem Restoration

Ecosystem restoration for the superb parrot on an abandoned railway line in Australia

According to the Society for Ecological Restoration, ecosystem restoration is the return of a damaged ecological system to a stable, healthy, and sustainable state, often together with associated ecosystem services.

Rationale

There are many reasons to restore ecosystems. Some include:

- Restoring natural capital such as drinkable water or wildlife populations.
- Mitigating climate change (e.g. through carbon sequestration)
- Helping threatened or endangered species
- Aesthetic reasons (Harris et al. 2006, Macdonald et al. 2002)
- Moral reasons: we have degraded, and in some cases destroyed, many ecosystems so it falls on us to 'fix' them.

There are considerable differences of opinion in how to set restoration goals and how to define their success. Some urge active restoration (e.g. eradicating invasive animals to allow the native ones to survive) and others who believe that protected areas should have the bare minimum of human interference. Ecosystem restoration has generated controversy, with skeptics who doubt that the benefits justify the economic investment or who point to failed restoration projects and question the feasibility of restoration altogether. It can be difficult to set restoration goals, in part because, as Anthony Bradshaw claims, "ecosystems are not static, but in a state of dynamic equilibrium.... [with restoration] we aim [for a] moving target."

Even though an ecosystem may not be returned to its original state, the functions of the ecosystem (especially ones that provide services to us) may be more valuable than its current configuration (Bradshaw 1987). One reason to consider ecosystem restoration is to mitigate climate change through activities such as afforestation. Afforestation involves replanting forests, which remove carbon dioxide from the air. Carbon dioxide is a leading cause of global warming (Speth, 2005) and capturing it would help alleviate climate change. Another example of a common driver of restoration projects in the United States is the legal framework of the Clean Water Act, which often requires mitigation for damage inflicted on aquatic systems by development or other activities.

Challenges in Restoration

Some view ecosystem restoration as impractical, in part because it sometimes fails. Hilderbrand et al. point out that many times uncertainty (about ecosystem functions, species relationships, and such) is not addressed, and that the time-scales set out for 'complete' restoration are unreasonably short. In other instances an ecosystem may be so degraded that abandonment (allowing an injured ecosystem to recover on its own) may be the wisest option (Holl, 2006). Local communities sometimes object to restorations that include the introduction of large predators or plants that require disturbance regimes such as regular fires (MacDonald et al. 2002). High economic costs can also be a perceived as a negative impact of the restoration process. Public opinion is very important in the feasibility of a restoration; if the public believes that the costs of restoration outweigh the benefits they will not support it (MacDonald et al. 2002). In these cases people might be ready to leave the ecosystem to recover on its own, which can sometimes occur relatively quickly (Holl, 2006).

Many failures have occurred in past restoration projects, many times because clear goals were not set out as the aim of the restoration. This may be because, as Peter Alpert says, "people may not [always] know how to manage natural systems effectively". Also many assumptions are made about myths of restoration such as the carbon copy, where a restoration plan, which worked in one area, is applied to another with the same results expected, but not realized (Hilderbrand et al. 2005).

Conservation Movement

Much attention has been given to preserving the natural characteristics of Hopetoun Falls, Australia, while allowing ample access for visitors.

The conservation movement, also known as nature conservation, is a political, environmental and a social movement that seeks to protect natural resources including animal and plant species as well as their habitat for the future.

The early conservation movement included fisheries and wildlife management, water, soil conservation and sustainable forestry. The contemporary conservation movement has broadened from the early movement's emphasis on use of sustainable yield of natural resources and preservation of wilderness areas to include preservation of biodiversity. Some say the conservation movement is part of the broader and more far-reaching environmental movement, while others argue that they differ both in ideology and practice. Chiefly in the United States, conservation is seen as differing from environmentalism in that it aims to preserve natural resources expressly for their continued sustainable use by humans. In other parts of the world conservation is used more broadly to include the setting aside of natural areas and the active protection of wildlife for their inherent value, as much as for any value they may have for humans.

History

Early History

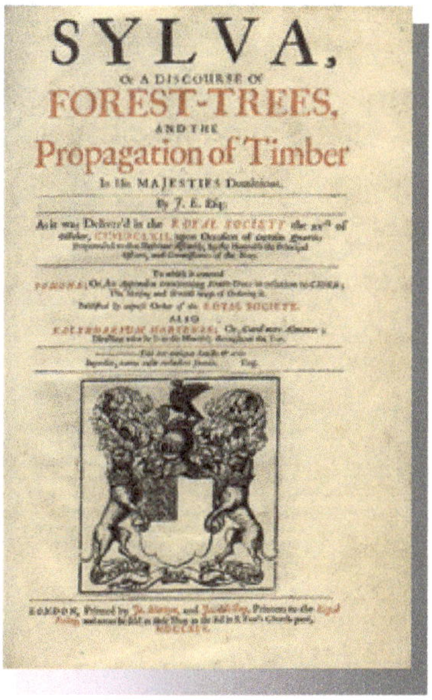

Sylva, or A Discourse of Forest-Trees and the Propagation of Timber in His Majesty's Dominions, title page of the first edition (1664).

The conservation movement can be traced back to John Evelyn's work *Sylva*, presented as a paper to the Royal Society in 1662. Published as a book two years later, it was one of the most highly influential texts on forestry ever published. Timber resources in England were becoming dangerously depleted at the time, and Evelyn advocated the importance of conserving the forests by managing the rate of depletion and ensuring that the cut down trees get replenished.

The field developed during the 18th century, especially in Prussia and France where scientific forestry methods were developed. These methods were first applied rigorously in British India from the early-19th century. The government was interested in the use of forest produce and began managing the forests with measures to reduce the risk of wildfire in order to protect the "household" of nature, as it was then termed. This early ecological idea was in order to preserve the growth of delicate teak trees, which was an important resource for the Royal Navy. Concerns over teak depletion were raised as early as 1799 and 1805 when the Navy was undergoing a massive expansion during the Napoleonic Wars; this pressure led to the first formal conservation Act, which prohibited the felling of small teak trees. The first forestry officer was appointed in 1806 to regulate and preserve the trees necessary for shipbuilding. This promising start received a setback in the 1820s and 30s, when laissez-faire economics and complaints from private landowners brought these early conservation attempts to an end.

Origins of the Modern Conservation Movement

Conservation was revived in the mid-19th century, with the first practical application of scientific conservation principles to the forests of India. The conservation ethic that began to evolve included three core principles: that human activity damaged the environment, that there was a civic duty to maintain the environment for future generations, and that scientific, empirically based methods should be applied to ensure this duty was carried out. Sir James Ranald Martin was prominent in promoting this ideology, publishing many medico-topographical reports that demonstrated the scale of damage wrought through large-scale deforestation and desiccation, and lobbying extensively for the institutionalization of forest conservation activities in British India through the establishment of Forest Departments. Edward Percy Stebbing warned of desertification of India. The Madras Board of Revenue started local conservation efforts in 1842, headed by Alexander Gibson, a professional botanist who systematically adopted a forest conservation program based on scientific principles. This was the first case of state management of forests in the world.

These local attempts gradually received more attention by the British government as the unregulated felling of trees continued unabated. In 1850, the British Association in Edinburgh formed a committee to study forest destruction at the behest of Dr. Hugh Cleghorn a pioneer in the nascent conservation movement.

Schlich, in the middle of the seated row, with students from the forestry school at Oxford, on a visit to the forests of Saxony in the year 1892.

He had become interested in forest conservation in Mysore in 1847 and gave several lectures at the Association on the failure of agriculture in India. These lectures influenced the government under Governor-General Lord Dalhousie to introduce the first permanent and large-scale forest conservation program in the world in 1855, a model that soon spread to other colonies, as well the United States. In the same year, Cleghorn organised the Madras Forest Department and in 1860 the Department banned the use shifting cultivation. Cleghorn's 1861 manual, *The forests and gardens of South India,* became the definitive work on the subject and was widely used by forest assistants in the subcontinent. In 1861, the Forest Department extended its remit into the Punjab.

Sir Dietrich Brandis, a German forester, joined the British service in 1856 as superintendent of the teak forests of Pegu division in eastern Burma. During that time Burma's teak forests were controlled by militant Karen tribals. He introduced the "taungya" system, in which Karen villagers provided labour for clearing, planting and weeding teak plantations. After seven years in Burma, Brandis was appointed Inspector General of Forests in India, a position he served in for 20 years. He formulated new forest legislation and helped establish research and training institutions. The Imperial Forest School at Dehradun was founded by him.

Germans were prominent in the forestry administration of British India. As well as Brandis, Berthold Ribbentrop and Sir William P.D. Schlich brought new methods to Indian conservation, the latter becoming the Inspector-General in 1883 after Brandis stepped down. Schlich helped to establish the journal *Indian Forester* in 1874, and became the founding director of the first forestry school in England at Cooper's Hill in 1885. He authored the five-volume *Manual of Forestry* (1889–96) on silviculture, forest management, forest protection, and forest utilisation, which became the standard and enduring textbook for forestry students.

Conservation in the United States

F. V. Hayden's map of Yellowstone National Park, 1871.

The American movement received its inspiration from 19th century works that exalted the inherent value of nature, quite apart from human usage. Author Henry David Thoreau (1817-1862) made key philosophical contributions that exalted nature. Thoreau was interested in peoples' relationship with nature and studied this by living close to nature in a simple life. He published his experiences in the book *Walden,* which argued that people should become intimately close with nature. The ideas of Sir Brandis, Sir William P.D. Schlich and Carl A. Schenck were also very influential - Gifford Pinchot, the first chief of the USDA Forest Service, relied heavily upon Brandis' advice for introducing professional forest management in the U.S. and on how to structure the Forest Service.

Both Conservationists and Preservationists appeared in political debates during the Progressive Era in the early 20th century. There were three main positions. The laissez-faire position held that owners of private property—including lumber and mining companies, should be allowed to do anything they wished for their property.

The conservationists, led by future President Theodore Roosevelt and his close ally George Bird Grinnell, were motivated by the wanton waste that was taking place at the hand of market forces, including logging and hunting. This practice resulted in placing a large number of North American game species on the edge of extinction. Roosevelt recognized that the laissez-faire approach of the U.S. Government was too wasteful and inefficient. In any case, they noted, most of the natural resources in the western states were already owned by the federal government. The best course of action, they argued, was a long-term plan devised by national experts to maximize the long-term economic benefits of natural resources. To accomplish the mission, Roosevelt and Grinnell formed the Boone and Crockett Club in 1887. The Club was made up of the best minds and influential men of the day. The Boone and Crockett Club's contingency of conservationists, scientists, politicians, and intellectuals became Roosevelt's closest advisers during his march to preserve wildlife and habitat across North America. Preservationists, led by John Muir (1838–1914), argued that the conservation policies were not strong enough to protect the interest of the natural world because they continued to focus on the natural world as a source of economic production.

The debate between conservation and preservation reached its peak in the public debates over the construction of California's Hetch Hetchy dam in Yosemite National Park which supplies the water supply of San Francisco. Muir, leading the Sierra Club, declared that the valley must be preserved for the sake of its beauty: "No holier temple has ever been consecrated by the heart of man."

President Roosevelt put conservationist issue high on the national agenda. He worked with all the major figures of the movement, especially his chief advisor on the matter, Gifford Pinchot and was deeply committed to conserving natural resources. He encouraged the Newlands Reclamation Act of 1902 to promote federal construction of dams to irrigate small farms and placed 230 million acres (360,000 mi^2 or 930,000 km^2) under federal protection. Roosevelt set aside more federal land for national parks and nature preserves than all of his predecessors combined.

A PRACTICAL FORESTER
(A subject that had attention all through Mr. Roosevelt's Presidency.)
From the *Pioneer Press* (St. Paul)

Roosevelt established the United States Forest Service, signed into law the creation of five national parks, and signed the year 1906 Antiquities Act, under which he proclaimed 18 new national monuments. He also established the first 51 bird reserves, four game preserves, and 150 national forests, including Shoshone National Forest, the nation's first. The area of the United States that he placed under public protection totals approximately 230,000,000 acres (930,000 km^2).

Gifford Pinchot had been appointed by McKinley as chief of Division of Forestry in the Department of Agriculture. In 1905, his department gained control of the national forest reserves. Pinchot promoted private use (for a fee) under federal supervision. In 1907, Roosevelt designated 16 million acres (65,000 km^2) of new national forests just minutes before a deadline.

In May 1908, Roosevelt sponsored the Conference of Governors held in the White House, with a focus on natural resources and their most efficient use. Roosevelt delivered the opening address: "Conservation as a National Duty.".

In 1903 Roosevelt toured the Yosemite Valley with John Muir, who had a very different view of conservation, and tried to minimize commercial use of water resources and forests. Working through the Sierra Club he founded, Muir succeeded in 1905 in having Congress transfer the Mariposa Grove and Yosemite Valley to the federal government. While Muir wanted nature preserved for its own sake, Roosevelt subscribed to Pinchot's formulation, "to make the forest produce the largest amount of whatever crop or service will be most useful, and keep on producing it for generation after generation of men and trees."

Theodore Roosevelt's view on conservationism remained dominant for decades; - Franklin D. Roosevelt authorised the building of many large-scale dams and water projects, as well as the expansion of the National Forest System to buy out sub-marginal farms. In 1937, the Pittman–Robertson Federal Aid in Wildlife Restoration Act was signed into law, providing funding for state agencies to carry out their conservation efforts.

Since 1970

Environmental reemerged on the national agenda in 1970, with Republican Richard Nixon playing a major role, especially with his creation of the Environmental Protection Agency. The debates over the public lands and environmental politics played a supporting role in the decline of liberalism and the rise of modern environmentalism. Although Americans consistently rank environmental issues as "important", polling data indicates that in the voting booth voters rank the environmental issues low relative to other political concerns.

The growth of the Republican party's political power in the inland West (apart from the Pacific coast) was facilitated by the rise of popular opposition to public lands reform. Successful Democrats in the inland West and Alaska typically take more conservative positions on environmental issues than Democrats from the Coastal states. Conservatives drew on new organizational networks of think tanks, industry groups, and citizen-oriented organizations, and they began to deploy new strategies that affirmed the rights of individuals to their property, protection of extraction rights, to hunt and recreate, and to pursue happiness unencumbered by the federal government at the expense of resource conservation.

Areas of Concern

Deforestation and overpopulation are issues affecting all regions of the world. The consequent destruction of wildlife habitat has prompted the creation of conservation groups in other countries, some founded by local hunters who have witnessed declining wildlife populations first hand. Also, it was highly important for the conservation movement to solve problems of living conditions in the cities and the overpopulation of such places.

Boreal Forest and the Arctic

The idea of incentive conservation is a modern one but its practice has clearly defended some of the sub Arctic wildernesses and the wildlife in those regions for thousands of years, especially by indigenous peoples such as the Evenk, Yakut, Sami, Inuit and Cree. The fur trade and hunting by these peoples have preserved these regions for thousands of years. Ironically, the pressure now upon them comes from non-renewable resources such as oil, sometimes to make synthetic clothing which is advocated as a humane substitute for fur. Similarly, in the case of the beaver, hunting and fur trade were thought to bring about the animal's demise, when in fact they were an integral part of its conservation. For many years children's books stated and still do, that the decline in the beaver population was due to the fur trade. In reality however, the decline in beaver numbers was because of habitat destruction and deforestation, as well as its continued persecution as a pest (it causes flooding). In Cree lands however, where the population valued the animal for meat and fur, it continued to thrive. The Inuit defend their relationship with the seal in response to outside critics.

Latin America (Bolivia)

The Izoceño-Guaraní of Santa Cruz Department, Bolivia is a tribe of hunters who were influential in establishing the Capitania del Alto y Bajo Isoso (CABI). CABI promotes economic growth and survival of the Izoceno people while discouraging the rapid destruction of habitat within Bolivia's Gran Chaco. They are responsible for the creation of the 34,000 square kilometre Kaa-Iya del Gran Chaco National Park and Integrated Management Area (KINP). The KINP protects the most biodiverse portion of the Gran Chaco, an ecoregion shared with Argentina, Paraguay and Brazil. In 1996, the Wildlife Conservation Society joined forces with CABI to institute wildlife and hunting monitoring programs in 23 Izoceño communities. The partnership combines traditional beliefs and local knowledge with the political and administrative tools needed to effectively manage habitats. The programs rely solely on voluntary participation by local hunters who perform self-monitoring techniques and keep records of their hunts. The information obtained by the hunters participating in the program has provided CABI with important data required to make educated decisions about the use of the land. Hunters have been willing participants in this program because of pride in their traditional activities, encouragement by their communities and expectations of benefits to the area.

Africa (Botswana)

In order to discourage illegal South African hunting parties and ensure future local use and sustainability, indigenous hunters in Botswana began lobbying for and implementing conservation practices in the 1960s. The Fauna Preservation Society of Ngamiland (FPS) was formed in 1962 by the husband and wife team: Robert Kay and June Kay, environmentalists working in conjunction with the Batawana tribes to preserve wildlife habitat.

The FPS promotes habitat conservation and provides local education for preservation of wildlife. Conservation initiatives were met with strong opposition from the Botswana government because of the monies tied to big-game hunting. In 1963, BaTawanga Chiefs and tribal hunter/adventurers in conjunction with the FPS founded Moremi National Park and Wildlife Refuge, the first area to be set aside by tribal people rather than governmental forces. Moremi National Park is home to a variety of wildlife, including lions, giraffes, elephants, buffalo, zebra, cheetahs and antelope, and covers an area of 3,000 square kilometers. Most of the groups involved with establishing this protected land were involved with hunting and were motivated by their personal observations of declining wildlife and habitat.

Conservation Biology

Conservation biology is the scientific study of nature and of Earth's biodiversity with the aim of protecting species, their habitats, and ecosystems from excessive rates of extinction and the erosion of biotic interactions. It is an interdisciplinary subject drawing on natural and social sciences, and the practice of natural resource management.

The conservation ethic is based on the findings of conservation biology.

Use of the Term

This term *conservation biology* was introduced as the title of a conference held at the University of California, San Diego in La Jolla, California in 1978 organized by biologists Bruce Wilcox and Michael E. Soulé. The meeting was prompted by the concern among scientists over tropical deforestation, disappearing species, eroding genetic diversity within species. The conference and proceedings that resulted sought to bridge a gap existing at the time between theory in ecology and population biology on the one hand and conservation policy and practice on the other. Conservation biology and the concept of biological diversity (biodiversity) emerged together, helping crystallize the modern era of conservation science and policy. The inherent multidisciplinary basis for conservation biology has led to the development of several new subdisciplines including conservation genetics, conservation social science, conservation behavior and conservation physiology.

Description

The rapid decline of established biological systems around the world means that conservation biology is often referred to as a "Discipline with a deadline". Conservation biology is tied closely to ecology in researching the dispersal, migration, demographics, effective population size, inbreeding depression, and minimum population viability of rare or endangered species. To better understand the restoration ecology of native plant and animal communities, the conservation biologist closely studies both their polytypic and monotypic habitats that are affected by a wide range of benign and hostile factors. Conservation biology is concerned with phenomena that affect the maintenance, loss, and restoration of biodiversity and the science of sustaining evolutionary processes that engender genetic, population, species, and ecosystem diversity. The concern stems from estimates suggesting that up to 50% of all species on the planet will disappear within the next 50 years, which has contributed to poverty, starvation, and will reset the course of evolution on this planet.

Conservation biologists research and educate on the trends and process of biodiversity loss, species extinctions, and the negative effect these are having on our capabilities to sustain the well-being of human society. Conservation biologists work in the field and office, in government, universities, non-profit organizations and industry. They are funded to research, monitor, and catalog every angle of the earth and its relation to society. The topics are diverse, because this is an interdisciplinary network with professional alliances in the biological as well as social sciences. Those dedicated to the cause and profession advocate for a global response to the current biodiversity crisis based on morals, ethics, and scientific reason. Organizations and citizens are responding to the biodiversity crisis through conservation action plans that direct research, monitoring, and education programs that engage concerns at local through global scales.

History

The conservation of natural indian resources is the fundamental problem. Unless we solve that problem, it will avail us little to solve all others.

– *Theodore Roosevelt*

Natural Resource Conservation

Conscious efforts to conserve and protect *global* biodiversity are a recent phenomenon. Natural resource conservation, however, has a history that extends prior to the age of conservation. Resource ethics grew out of necessity through direct relations with nature. Regulation or communal restraint became necessary to prevent selfish motives from taking more than could be locally sustained, therefore compromising the long-term supply for the rest of the community. This social dilemma with respect to natural resource management is often called the "Tragedy of the Commons".

From this principle, conservation biologists can trace communal resource based ethics throughout cultures as a solution to communal resource conflict. For example, the Alaskan Tlingit peoples and the Haida of the Pacific Northwest had resource boundaries, rules, and restrictions among clans with respect to the fishing of sockeye salmon. These rules were guided by clan elders who knew lifelong details of each river and stream they managed. There are numerous examples in history where cultures have followed rules, rituals, and organized practice with respect to communal natural resource management.

Conservation ethics are also found in early religious and philosophical writings. There are examples in the Tao, Shinto, Hindu, Islamic and Buddhist traditions. In Greek philosophy, Plato lamented about pasture land degradation: "What is left now is, so to say, the skeleton of a body wasted by disease; the rich, soft soil has been carried off and only the bare framework of the district left." In the bible, through Moses, God commanded to let the land rest from cultivation every seventh year. Before the 18th century, however, much of European culture considered it a pagan view to admire nature. Wilderness was denigrated while agricultural development was praised. However, as early as AD 680 a wildlife sanctuary was founded on the Farne Islands by St Cuthbert in response to his religious beliefs.

Early Naturalists

White gyrfalcons drawn by John James Audubon

Natural history was a major preoccupation in the 18th century, with grand expeditions and the opening of popular public displays in Europe and North America. By 1900 there were 150 natural history museums in Germany, 250 in Great Britain, 250 in the United States, and 300 in France. Preservationist or conservationist sentiments are a development of the late 18th to early 20th centuries.

By the early 19th century biogeography was ignited through the efforts of Alexander von Humboldt, Charles Lyell and Charles Darwin. The 19th-century fascination with natural history engendered a fervor to be the first to collect rare specimens with the goal of doing so before they became extinct by other such collectors. Although the work of many 18th and 19th century naturalists were to inspire nature enthusiasts and conservation organizations, their writings, by modern standards, showed insensitivity towards conservation as they would kill hundreds of specimens for their collections.

Conservation Movement

The modern roots of conservation biology can be found in the late 18th-century Enlightenment period particularly in England and Scotland. A number of thinkers, among them notably Lord Monboddo, described the importance of "preserving nature"; much of this early emphasis had its origins in Christian theology.

Scientific conservation principles were first practically applied to the forests of British India. The conservation ethic that began to evolve included three core principles:

that human activity damaged the environment, that there was a civic duty to maintain the environment for future generations, and that scientific, empirically based methods should be applied to ensure this duty was carried out. Sir James Ranald Martin was prominent in promoting this ideology, publishing many medico-topographical reports that demonstrated the scale of damage wrought through large-scale deforestation and desiccation, and lobbying extensively for the institutionalization of forest conservation activities in British India through the establishment of Forest Departments.

The Madras Board of Revenue started local conservation efforts in 1842, headed by Alexander Gibson, a professional botanist who systematically adopted a forest conservation program based on scientific principles. This was the first case of state conservation management of forests in the world. Governor-General Lord Dalhousie introduced the first permanent and large-scale forest conservation program in the world in 1855, a model that soon spread to other colonies, as well the United States, where Yellowstone National Park was opened in 1872 as the world's first national park.

The term *conservation* came into widespread use in the late 19th century and referred to the management, mainly for economic reasons, of such natural resources as timber, fish, game, topsoil, pastureland, and minerals. In addition it referred to the preservation of forests (forestry), wildlife (wildlife refuge), parkland, wilderness, and watersheds. This period also saw the passage of the first conservation legislation and the establishment of the first nature conservation societies. The Sea Birds Preservation Act of 1869 was passed in Britain as the first nature protection law in the world after extensive lobbying from the Association for the Protection of Seabirds and the respected ornithologist Alfred Newton. Newton was also instrumental in the passage of the first Game laws from 1872, which protected animals during their breeding season so as to prevent the stock from being brought close to extinction.

One of the first conservation societies was the Royal Society for the Protection of Birds, founded in 1889 in Manchester as a protest group campaigning against the use of great crested grebe and kittiwake skins and feathers in fur clothing. Originally known as "the Plumage League". the group gained popularity and eventually amalgamated with the Fur and Feather League in Croydon, and formed the RSPB. The National Trust formed in 1895 with the manifesto to "...promote the permanent preservation, for the benefit of the nation, of lands, ...to preserve (so far practicable) their natural aspect."

In the United States, the Forest Reserve Act of 1891 gave the President power to set aside forest reserves from the land in the public domain. John Muir founded the Sierra Club in 1892, and the New York Zoological Society was set up in 1895. A series of national forests and preserves were established by Theodore Roosevelt from 1901 to 1909. The 1916 National Parks Act, included a 'use without impairment' clause, sought by John Muir, which eventually resulted in the removal of a proposal to build a dam in Dinosaur National Monument in 1959.

Roosevelt and Muir on Glacier Point in Yosemite National Park.

In the 20th century, Canadian civil servants, including Charles Gordon Hewitt and James Harkin spearheaded the movement toward wildlife conservation.

Global Conservation Efforts

In the mid-20th century, efforts arose to target individual species for conservation, notably efforts in big cat conservation in South America led by the New York Zoological Society. In the early 20th century the New York Zoological Society was instrumental in developing concepts of establishing preserves for particular species and conducting the necessary conservation studies to determine the suitability of locations that are most appropriate as conservation priorities; the work of Henry Fairfield Osborn Jr., Carl E. Akeley, Archie Carr and Archie Carr III is notable in this era. Akeley for example, having led expeditions to the Virunga Mountains and observed the mountain gorilla in the wild, became convinced that the species and the area were conservation priorities. He was instrumental in persuading Albert I of Belgium to act in defense of the mountain gorilla and establish Albert National Park (since renamed Virunga National Park) in what is now Democratic Republic of Congo.

By the 1970s, led primarily by work in the United States under the Endangered Species Act along with the Species at Risk Act (SARA) of Canada, Biodiversity Action Plans developed in Australia, Sweden, the United Kingdom, hundreds of species specific protection plans ensued. Notably the United Nations acted to conserve sites of outstanding cultural or natural importance to the common heritage of mankind. The programme was adopted by the General Conference of UNESCO in 1972. As of 2006, a total of 830 sites are listed: 644 cultural, 162 natural. The first country to pursue aggressive biological conservation through national legislation was the United States, which passed back to back legislation in the Endangered Species Act (1966) and National Environmental Policy Act (1970), which together injected major funding and protection measures to large-scale habitat protection and threatened species research. Other conservation developments, however, have taken hold throughout the world. India, for example, passed the Wildlife Protection Act of 1972 .

In 1980 a significant development was the emergence of the urban conservation movement. A local organization was established in Birmingham, UK, a development followed in rapid succession in cities across the UK, then overseas. Although perceived as a grassroots movement, its early development was driven by academic research into urban wildlife. Initially perceived as radical, the movement's view of conservation being inextricably linked with other human activity has now become mainstream in conservation thought. Considerable research effort is now directed at urban conservation biology. The Society for Conservation Biology originated in 1985.

By 1992 most of the countries of the world had become committed to the principles of conservation of biological diversity with the Convention on Biological Diversity; subsequently many countries began programmes of Biodiversity Action Plans to identify and conserve threatened species within their borders, as well as protect associated habitats. The late 1990s saw increasing professionalism in the sector, with the maturing of organisations such as the Institute of Ecology and Environmental Management and the Society for the Environment.

Since 2000 the concept of landscape scale conservation has risen to prominence, with less emphasis being given to single-species or even single-habitat focused actions. Instead an ecosystem approach is advocated by most mainstream conservationists, although concerns have been expressed by those working to protect some high-profile species.

Ecology has clarified the workings of the biosphere; i.e., the complex interrelationships among humans, other species, and the physical environment. The burgeoning human population and associated agriculture, industry, and the ensuing pollution, have demonstrated how easily ecological relationships can be disrupted.

> “ The last word in ignorance is the man who says of an animal or plant: “What good is it?” If the land mechanism as a whole is good, then every part is good, whether we understand it or not. If the biota, in the course of aeons, has built something we like but do not understand, then who but a fool would discard seemingly useless parts? To keep every cog and wheel is the first precaution of intelligent tinkering. ”
>
> —*Aldo Leopold, A Sand County Almanac*

Concepts and Foundations

Measuring Extinction Rates

Extinction rates are measured in a variety of ways. Conservation biologists measure and apply statistical measures of fossil records, rates of habitat loss, and a multitude of other variables such as loss of biodiversity as a function of the rate of habitat loss and site occupancy to obtain such estimates. *The Theory of Island Biogeography* is possibly the most significant contribution toward the scientific understanding of both the process and how to measure the rate of species extinction. The current background extinction rate is estimated to be one species every few years.

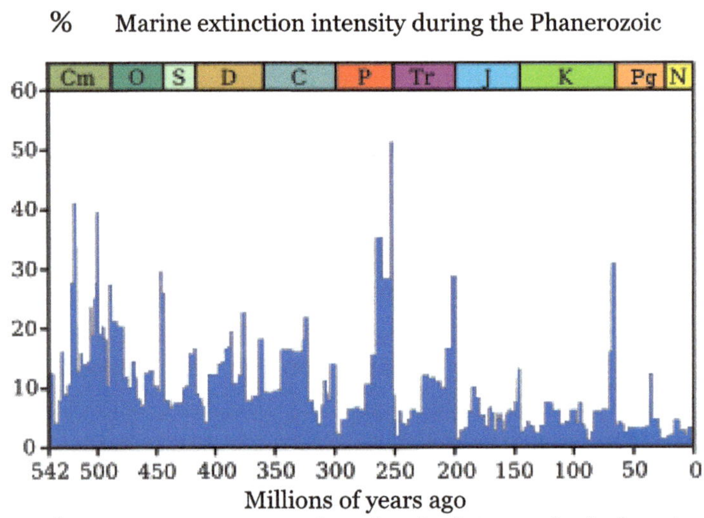

The blue graph shows the apparent *percentage* (not the absolute number) of marine animal genera becoming extinct during any given time interval. It does not represent all marine species, just those that are readily fossilized.

The measure of ongoing species loss is made more complex by the fact that most of the Earth's species have not been described or evaluated. Estimates vary greatly on how many species actually exist (estimated range: 3,600,000-111,700,000) to how many have received a species binomial (estimated range: 1.5-8 million). Less than 1% of all species that have been described have been studied beyond simply noting its existence. From these figures, the IUCN reports that 23% of vertebrates, 5% of invertebrates and 70% of plants that have been evaluated are designated as endangered or threatened. Better knowledge is being constructed by The Plant List for actual numbers of species.

Systematic Conservation Planning

Systematic conservation planning is an effective way to seek and identify efficient and effective types of reserve design to capture or sustain the highest priority biodiversity values and to work with communities in support of local ecosystems. Margules and Pressey identify six interlinked stages in the systematic planning approach:

1. Compile data on the biodiversity of the planning region
2. Identify conservation goals for the planning region

3. Review existing conservation areas
4. Select additional conservation areas
5. Implement conservation actions
6. Maintain the required values of conservation areas

Conservation biologists regularly prepare detailed conservation plans for grant proposals or to effectively coordinate their plan of action and to identify best management practices (e.g.). Systematic strategies generally employ the services of Geographic Information Systems to assist in the decision making process.

Conservation Physiology: a Mechanistic Approach to Conservation

Conservation Physiology was defined by Steven J. Cooke and colleagues as: 'An integrative scientific discipline applying physiological concepts, tools, and knowledge to characterizing biological diversity and its ecological implications; understanding and predicting how organisms, populations, and ecosystems respond to environmental change and stressors; and solving conservation problems across the broad range of taxa (i.e. including microbes, plants, and animals). Physiology is considered in the broadest possible terms to include functional and mechanistic responses at all scales, and conservation includes the development and refinement of strategies to rebuild populations, restore ecosystems, inform conservation policy, generate decision-support tools, and manage natural resources.' Conservation physiology is particularly relevant to practitioners in that it has the potential to generate cause-and-effect relationships and reveal the factors that contribute to population declines.

Conservation Biology as a Profession

The Society for Conservation Biology is a global community of conservation professionals dedicated to advancing the science and practice of conserving biodiversity. Conservation biology as a discipline reaches beyond biology, into subjects such as philosophy, law, economics, humanities, arts, anthropology, and education. Within biology, conservation genetics and evolution are immense fields unto themselves, but these disciplines are of prime importance to the practice and profession of conservation biology.

Is conservation biology an objective science when biologists advocate for an inherent value in nature? Do conservationists introduce bias when they support policies using qualitative description, such as habitat *degradation*, or *healthy* ecosystems? As all scientists hold values, so do conservation biologists. Conservation biologists advocate for reasoned and sensible management of natural resources and do so with a disclosed combination of science, reason, logic, and values in their conservation management plans. This sort of advocacy is similar to the medical profession advocating for healthy lifestyle options, both are beneficial to human well-being yet remain scientific in their approach.

There is a movement in conservation biology suggesting a new form of leadership is needed to mobilize conservation biology into a more effective discipline that is able to communicate the full scope of the problem to society at large. The movement proposes an adaptive leadership approach that parallels an adaptive management approach. The concept is based on a new philosophy or leadership theory steering away from historical notions of power, authority, and dominance. Adaptive conservation leadership is reflective and more equitable as it applies to any member of society who can mobilize others toward meaningful change using communication techniques that are inspiring, purposeful, and collegial. Adaptive conservation leadership and mentoring programs are being implemented by conservation biologists through organizations such as the Aldo Leopold Leadership Program

Approaches

Conservation may be classified as either in-situ conservation, which is protecting an endangered species in its natural habitat, or ex-situ conservation, which occurs outside the natural habitat. In-situ conservation involves protecting or cleaning up the habitat itself which may include a great deal of environmental preservation, or by defending the species from predators. Ex-situ conservation may be used on some or all of the population, when in-situ conservation is too difficult, or impossible.

Also, non-interference may be used, which is termed a preservationist method. Preservationists advocate for giving areas of nature and species a protected existence that halts interference from the humans. In this regard, conservationists differ from preservationists in the social dimension, as conservation biology engages society and seeks equitable solutions for both society and ecosystems.

Some preservationists emphasize the potential of biodiversity in a world without humans

"Animals have not yet invaded 2/3 of Earth's habitats, and it could be that without human influence the diversity of tetrapods will continue to increase in an exponential fashion."

—*Sahney et al.*

Ethics and Values

Conservation biologists are interdisciplinary researchers that practice ethics in the biological and social sciences. Chan states that conservationists must advocate for biodiversity and can do so in a scientifically ethical manner by not promoting simultaneous advocacy against other competing values. A conservationist researches biodiversity and reasons through a Resource Conservation Ethic , which identify what measures will deliver "the greatest good for the greatest number of people for the longest time."

Some conservation biologists argue that nature has an intrinsic value that is independent of anthropocentric usefulness or utilitarianism. Intrinsic value advocates that a gene, or species, be valued because they have a utility for the ecosystems they sustain. Aldo Leopold was a classical thinker and writer on such conservation ethics whose philosophy, ethics and writings are still valued and revisited by modern conservation biologists. His writing is often required reading for those in the profession.

Conservation Priorities

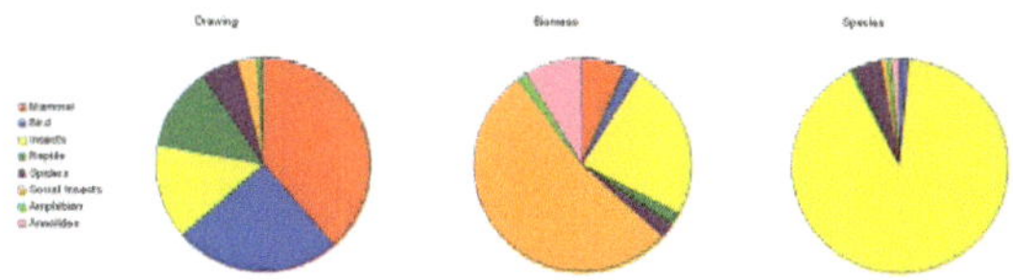

A pie chart image showing the relative biomass representation in a rain forest through a summary of children's perceptions from drawings and artwork (left), through a scientific estimate of actual biomass (middle), and by a measure of biodiversity (right). Notice that the biomass of social insects (middle) far outweighs the number of species (right).

The International Union for the Conservation of Nature (IUCN) has organized a global assortment of scientists and research stations across the planet to monitor the changing state of nature in an effort to tackle the extinction crisis. The IUCN provides annual updates on the status of species conservation through its Red List. The IUCN Red List serves as an international conservation tool to identify those species most in need of conservation attention and by providing a global index on the status of biodiversity. More than the dramatic rates of species loss, however, conservation scientists note that the sixth mass extinction is a biodiversity crisis requiring far more action than a priority focus on rare, endemic or endangered species. Concerns for biodiversity loss covers a broader conservation mandate that looks at ecological processes, such as migration, and a holistic examination of biodiversity at levels beyond the species, including genetic, population and ecosystem diversity. Extensive, systematic, and rapid rates of biodiversity loss threatens the sustained well-being of humanity by limiting supply of ecosystem services that are otherwise regenerated by the complex and evolving holistic network of genetic and ecosystem diversity. While the conservation status of species is employed extensively in conservation management, some scientists highlight that it is the common species that are the primary source of exploitation and habitat alteration by humanity. Moreover, common species are often undervalued despite their role as the primary source of ecosystem services.

While most in the community of conservation science "stress the importance" of sustaining biodiversity, there is debate on how to prioritize genes, species, or ecosystems, which are all components of biodiversity (e.g. Bowen, 1999). While the predominant approach to date has been to focus efforts on endangered species by conserving *biodiversity hotspots*, some scientists (e.g) and conservation organizations, such as the Nature Conservancy, argue that it is more cost-effective, logical, and socially relevant to invest in *biodiversity coldspots*. The costs of discovering, naming, and mapping out the distribution of every species, they argue, is an ill-advised conservation venture. They reason it is better to understand the significance of the ecological roles of species.

Biodiversity hotspots and coldspots are a way of recognizing that the spatial concentration of genes, species, and ecosystems is not uniformly distributed on the Earth's surface. For example, "[...] 44% of all species of vascular plants and 35% of all species in four vertebrate groups are confined to 25 hotspots comprising only 1.4% of the land surface of the Earth."

Those arguing in favor of setting priorities for coldspots point out that there are other measures to consider beyond biodiversity. They point out that emphasizing hotspots downplays the importance of the social and ecological connections to vast areas of the Earth's ecosystems where biomass, not biodiversity, reigns supreme. It is estimated that 36% of the Earth's surface, encompassing 38.9% of the worlds vertebrates, lacks the endemic species to qualify as biodiversity hotspot. Moreover, measures show that maximizing protections for biodiversity does not capture ecosystem services any better than targeting randomly chosen regions. Population level biodiversity (i.e. coldspots) are disappearing at a rate that is ten times that at the species level. The level of importance in addressing biomass versus endemism as a concern for conservation biology is highlighted in literature measuring the level of threat to global ecosystem carbon stocks that do not necessarily reside in areas of endemism. A hotspot priority approach would not invest so heavily in places such as steppes, the Serengeti, the Arctic, or taiga. These areas contribute a great abundance of population (not species) level biodiversity and ecosystem services, including cultural value and planetary nutrient cycling.

Those in favor of the hotspot approach point out that species are irreplaceable components of the global ecosystem, they are concentrated in places that are most threatened, and should therefore receive maximal strategic protections. The IUCN Red List categories, is an example of the hotspot conservation approach in action; species that are not rare or endemic are listed the least concern and their Wikipedia articles tend to be ranked low on the importance scale. This is a hotspot approach because the priority is set to target species level concerns over population level or biomass. Species richness and genetic biodiversity contributes to and engenders ecosystem stability, ecosystem processes, evolutionary adaptability, and biomass. Both sides agree, however, that conserving biodiversity is necessary to reduce the extinction rate and identify an inherent value in nature; the debate hinges on how to prioritize limited conservation resources in the most cost-ef-fective way.

Economic Values and Natural Capital

Tadrart Acacus desert in western Libya, part of the Sahara.

Conservation biologists have started to collaborate with leading global economists to determine how to measure the wealth and services of nature and to make these values apparent in global market transactions. This system of accounting is called *natural capital* and would, for example, register the value of an ecosystem before it is cleared to make way for development. The WWF publishes its *Living Planet Report* and provides a global index of biodiversity by monitoring approximately 5,000 populations in 1,686 species of vertebrate (mammals, birds, fish, reptiles, and amphibians) and report on the trends in much the same way that the stock market is tracked.

This method of measuring the global economic benefit of nature has been endorsed by the G8+5 leaders and the European Commission. Nature sustains many ecosystem services that benefit humanity. Many of the earths ecosystem services are public goods without a market and therefore no price or value. When the *stock market* registers a financial crisis, traders on Wall Street are not in the business of trading stocks for much of the planet's living natural capital stored in ecosystems. There is no natural stock market with investment portfolios into sea horses, amphibians, insects, and other creatures that provide a sustainable supply of ecosystem services that are valuable to society. The ecological footprint of society has exceeded the bio-regenerative capacity limits of the planet's ecosystems by about 30 percent, which is the same percentage of vertebrate populations that have registered decline from 1970 through 2005.

The ecological credit crunch is a global challenge. The *Living Planet Report 2008* tells us that more than three-quarters of the world's people live in nations that are ecological debtors – their national consumption has outstripped their country's biocapacity. Thus, most of us are propping up our current lifestyles, and our economic growth, by drawing (and increasingly overdrawing) upon the ecological capital of other parts of the world.

WWF Living Planet Report

The inherent natural economy plays an essential role in sustaining humanity, including the regulation of global atmospheric chemistry, pollinating crops, pest control, cycling soil nutrients, purifying our water supply, supplying medicines and health benefits, and unquantifiable quality of life improvements. There is a relationship, a correlation, between markets and natural capital, and social income inequity and biodiversity loss. This means that there are greater rates of biodiversity loss in places where the inequity of wealth is greatest

Although a direct market comparison of natural capital is likely insufficient in terms of human value, one measure of ecosystem services suggests the contribution amounts to trillions of dollars yearly. For example, one segment of North American forests has been assigned an annual value of 250 billion dollars; as another example, honey-bee pollination is estimated to provide between 10 and 18 billion dollars of value yearly. The value of ecosystem services on one New Zealand island has been imputed to be as great as the GDP of that region. This planetary wealth is being lost at an incredible rate as the demands of human society is exceeding the bio-regenerative capacity of the Earth. While biodiversity and ecosystems are resilient, the danger of losing them is that humans cannot recreate many ecosystem functions through technological innovation.

Strategic Species Concepts

Keystone Species

Some species, called a *keystone species*, form a central supporting hub in the ecosystem. The loss of such a species results in a collapse in ecosystem function, as well as the loss of coexisting species. The importance of a keystone species was shown by the extinction of the Steller's sea cow (*Hydrodamalis gigas*) through its interaction with sea otters, sea urchins, and kelp. Kelp beds grow and form nurseries in shallow waters to shelter creatures that support the food chain. Sea urchins feed on kelp, while sea otters feed on sea urchins. With the rapid decline of sea otters due to overhunting, sea urchin populations grazed unrestricted on the kelp beds and the ecosystem collapsed. Left unchecked, the urchins destroyed the shallow water kelp communities that supported the Steller's sea cow's diet and hastened their demise. The sea otter was thought to be a keystone species because the coexistence of many ecological associates in the kelp beds relied upon otters for their survival. However this was later questioned by Turvey and Risley, who showed that hunting alone would have driven the Steller's sea cow extinct.

Indicator Species

An *indicator species* has a narrow set of ecological requirements, therefore they become useful targets for observing the health of an ecosystem. Some animals, such as amphibians with their semi-permeable skin and linkages to wetlands, have an acute sensitivity to environmental harm and thus may serve as a *miner's canary*. Indicator species are monitored in an effort to capture environmental degradation through pollution or some other link to proximate human activities. Monitoring an indicator species is a measure to determine if there is a significant environmental impact that can serve to advise or modify practice, such as through different forest silviculture treatments and management scenarios, or to measure the degree of harm that a pesticide may impart on the health of an ecosystem.

Government regulators, consultants, or NGOs regularly monitor indicator species, however, there are limitations coupled with many practical considerations that must be followed for the approach to be effective. It is generally recommended that multiple indicators (genes, populations, species, communities, and landscape) be monitored for effective conservation measurement that prevents harm to the complex, and often unpredictable, response from ecosystem dynamics (Noss, 1997).

Umbrella and Flagship Species

An example of an *umbrella species* is the monarch butterfly, because of its lengthy migrations and aesthetic value. The monarch migrates across North America, covering multiple ecosystems and so requires a large area to exist. Any protections afforded to the monarch butterfly will at the same time umbrella many other species and habitats. An umbrella species is often used as *flagship species*, which are species, such as the giant panda, the blue whale, the tiger, the mountain gorilla and the monarch butterfly, that capture the public's attention and attract support for conservation measures.

Context and Trends

Conservation biologists study trends and process from the paleontological past to the ecological present as they gain an understanding of the context related to species extinction. It is generally accepted that there have been five major global mass extinctions that register in Earth's history. These include: the Ordovician (440 mya), Devonian (370 mya), Permian–Triassic (245 mya), Triassic–Jurassic (200 mya), and Cretaceous–Paleogene extinction event (66 mya) extinction spasms. Within the last 10,000 years, human influence over the Earth's ecosystems has been so extensive that scientists have difficulty estimating the number of species lost; that is to say the rates of deforestation, reef destruction, wetland draining and other human acts are proceeding much faster than human assessment of species. The latest *Living Planet Report* by the World Wide Fund for Nature estimates that we have exceeded the bio-regenerative capacity of the planet, requiring 1.5 Earths to support the demands placed on our natural resources.

Holocene Extinction

Conservation biologists are dealing with and have published evidence from all corners of the planet indicating that humanity may be causing the sixth and fastest planetary extinction event. It has been suggested that we are living in an era of unprecedented numbers of species extinctions, also known as the Holocene extinction event. The global extinction rate may be approximately 100,000 times higher than the natural background extinction rate. It is estimated that two-thirds of all mammal genera and one-half of all mammal species weighing at least 44 kilograms (97 lb) have gone extinct in the last 50,000 years. The Global Amphibian Assessment reports that amphibians are declining on a global scale faster than any other vertebrate group, with over 32% of all surviving species being threatened with extinction. The surviving populations are in continual decline in 43% of those that are threatened. Since the mid-1980s the actual rates of extinction have exceeded 211 times rates measured from the fossil record. However, "The current amphibian extinction rate may range from 25,039 to 45,474 times the background extinction rate for amphibians." The global extinction trend occurs in every major vertebrate group that is being monitored. For example, 23% of all mammals and 12% of all birds are Red Listed by the International Union for Conservation of Nature (IUCN), meaning they too are threatened with extinction.

An art scape image showing the relative importance of animals in a rain forest through a summary of (a) child's perception compared with (b) a scientific estimate of the importance. The size of the animal represents its importance. The child's mental image places importance on big cats, birds, butterflies, and then reptiles versus the actual dominance of social insects (such as ants).

Status of Oceans and Reefs

Global assessments of coral reefs of the world continue to report drastic and rapid rates of decline. By 2000, 27% of the world's coral reef ecosystems had effectively collapsed. The largest period of decline occurred in a dramatic "bleaching" event in 1998, where approximately 16% of all the coral reefs in the world disappeared in less than a year. *Coral bleaching* is caused by a mixture of environmental stresses, including increases in ocean temperatures and acidity, causing both the release of symbiotic algae and death of corals. Decline and extinction risk in coral reef biodiversity has risen dramatically in the past ten years. The loss of coral reefs, which are predicted to go extinct in the next century, threatens the balance of global biodiversity, will have huge economic impacts, and endangers food security for hundreds of millions of people. Conservation biology plays an important role in international agreements covering the world's oceans (and other issues pertaining to biodiversity, e.g.).

These predictions will undoubtedly appear extreme, but it is difficult to imagine how such changes will not come to pass without fundamental changes in human behavior.

J.B. Jackson

The oceans are threatened by acidification due to an increase in CO_2 levels. This is a most serious threat to societies relying heavily upon oceanic natural resources. A concern is that the majority of all marine species will not be able to evolve or acclimate in response to the changes in the ocean chemistry.

The prospects of averting mass extinction seems unlikely when "[...] 90% of all of the large (average approximately ≥50 kg), open ocean tuna, billfishes, and sharks in the ocean" are reportedly gone. Given the scientific review of current trends, the ocean is predicted to have few surviving multi-cellular organisms with only microbes left to dominate marine ecosystems.

Groups other than Vertebrates

Serious concerns also being raised about taxonomic groups that do not receive the same degree of social attention or attract funds as the vertebrates. These include fungal (including lichen-forming species), invertebrate (particularly insect) and plant communities where the vast majority of biodiversity is represented. Conservation of fungi and conservation of insects, in particular, are both of pivotal importance for conservation biology. As mycorrhizal symbionts, and as decomposers and recyclers, fungi are essential for sustainability of forests. The value of insects in the biosphere is enormous because they outnumber all other living groups in measure of species richness. The greatest bulk of biomass on land is found in plants, which is sustained by insect relations. This great ecological value of insects is countered by a society that often reacts negatively toward these aesthetically 'unpleasant' creatures.

One area of concern in the insect world that has caught the public eye is the mysterious case of missing honey bees (*Apis mellifera*). Honey bees provide an indispensable ecological services through their acts of pollination supporting a huge variety of agriculture crops. The sudden disappearance of bees leaving empty hives or colony collapse disorder (CCD) is not uncommon. However, in 16-month period from 2006 through 2007, 29% of 577 beekeepers across the United States reported CCD losses in up to 76% of their colonies. This sudden demographic loss in bee numbers is placing a strain on the agricultural sector. The cause behind the massive declines is puzzling scientists. Pests, pesticides, and global warming are all being considered as possible causes.

Another highlight that links conservation biology to insects, forests, and climate change is the mountain pine beetle (*Dendroctonus ponderosae*) epidemic of British Columbia, Canada, which has infested 470,000 km² (180,000 sq mi) of forested land since 1999. An action plan has been prepared by the Government of British Columbia to address this problem.

This impact [*pine beetle epidemic*] converted the forest from a small net carbon sink to a large net carbon source both during and immediately after the outbreak. In the worst year, the impacts resulting from the beetle outbreak in British Columbia were equivalent to 75% of the average annual direct forest fire emissions from all of Canada during 1959–1999.

—*Kurz et al.*

Conservation Biology of Parasites

A large proportion of parasite species are threatened by extinction. A few of them are being eradicated as pests of humans or domestic animals, however, most of them are harmless. Threats include the decline or fragmentation of host populations, or the extinction of host species.

Threats to Biodiversity

Many of the threats to biodiversity, including disease and climate change, are reaching inside borders of protected areas, leaving them 'not-so protected' (e.g. Yellowstone National Park). Climate change, for example, is often cited as a serious threat in this regard, because there is a feedback loop between species extinction and the release of carbon dioxide into the atmosphere. Ecosystems store and cycle large amounts of carbon which regulates global conditions. The effects of global warming add a catastrophic threat toward a mass extinction of global biological diversity. The extinction threat is estimated to range from 15 to 37 percent of all species by 2050, or 50 percent of all species over the next 50 years.

Some of the most significant threats to biodiversity and ecosystem processes include climate change, mass agriculture, deforestation, overgrazing, slash-and-burn agriculture, urban development, wildlife trade, light pollution and pesticide use. Habitat fragmentation poses one of the more difficult challenges, because the global network of protected areas only covers 11.5% of the Earth's surface. Roads are one cause of fragmentation, as well as a direct sourcc of mortality for many types of animals, but they can also have some beneficial effects. A significant consequence of fragmentation and lack of linked protected areas is the reduction of animal migration on a global scale. Considering that billions of tonnes of biomass are responsible for nutrient cycling across the earth, the reduction of migration is a serious matter for conservation biology.

Human activities are associated directly or indirectly with nearly every aspect of the current extinction spasm.

Wake and Vredenburg

These figures do not imply, however, that human activities must necessarily cause irreparable harm to the biosphere. With conservation management and planning for biodiversity at all levels, from genes to ecosystems, there are examples where humans mutually coexist in a sustainable way with nature. Even with the current threats to biodiversity there are always ways we can improve the current condition and start anew.

Disturbance (Ecology)

Damages of storm Kyrill in Wittgenstein, Germany.

In biology, a disturbance is a temporary change in environmental conditions that causes a pronounced change in an ecosystem. Disturbances often act quickly and with great effect, to alter the physical structure or arrangement of biotic and abiotic elements. Major ecological disturbances may include fires, flooding, windstorms, insect outbreaks and trampling. Earthquakes, various types of volcanic eruptions, tsunami, firestorms, impact events, climate change, and the devastating effects of human impact on the environment (anthropogenic disturbances) such as clearcutting, forest clearing and the introduction of invasive species can be considered major disturbances. Disturbance forces can have profound immediate effects on ecosystems and can, accordingly, greatly alter the natural community. Because of these and the impacts on populations, these effects can continue for an extended period of time.

Criteria

Conditions under which natural disturbances occur are influenced mainly by climate, weather, and location. Fire disturbances will only occur in areas where there is low precipitation, some form of ignition (typically lightning), and enough flammable biomass to allow fire to spread. Conditions often occur as part of a cycle and disturbances may be periodic. Other disturbances, such as those caused by humans, invasive species or impact events, can occur anywhere and are not necessarily cyclic. Extinction vortices may result in multiple disturbances or a greater frequency of a single disturbance.

Cyclic Disturbance

Often, when disturbances occur naturally, they provide conditions that favor the success of different species over pre-disturbance organisms. This can be attributed to physical changes in the biotic and abiotic conditions of an ecosystem. Because of this, a disturbance force can change an ecosystem for significantly longer than the period over which the immediate effects persist. However, in the absence of further disturbance forces, many ecosystems will trend back toward pre-disturbance conditions. Such alteration, accompanied by changes in the abundance of different species over time, is called ecological succession. Succession often leads to conditions that will once again predispose an ecosystem to disturbance.

Pine forests in the western North America provide a good example of such a cycle involving insect outbreaks. The mountain pine beetle (*Dendroctonus ponderosae*) play an important role in limiting pine trees like lodgepole pine in forests of western North America. In 2004 the beetles affected more than 90,000 square kilometres. The beetles exist in endemic and epidemic phases. During epidemic phases swarms of beetles kill large numbers of old pines. This mortality creates openings in the forest for new vegetation. Spruce, fir, and younger pines, which are unaffected by the beetles, thrive in canopy openings. Eventually pines grow into the canopy and replace those lost. Younger pines are often able to ward off beetle attacks but, as they grow older, pines become less vigorous and more susceptible to infestation. This cycle of death and re-growth creates a temporal mosaic of pines in the forest. Similar cycles occur in association with other disturbances such as fire and windthrow.

Compound Disturbances

When multiple disturbance events affect the same location in quick succession, this often results in a "compound disturbance," an event which, due to the combination of forces, creates a new situation which is more than the sum of its parts. For example, windstorms followed by fire can create fire temperatures and durations that are not expected in even severe wildfires, and may have surprising effects on post-fire succession.

Species Adapted to Disturbance

A disturbance may change a forests significantly. Afterwards, the forest floor is often littered with dead material. This decaying matter and abundant sunlight promote an abundance of new growth. In the case of forest fires a portion of the nutrients previously held in plant biomass is returned quickly to the soil as biomass burns. Many plants and animals benefit from disturbance conditions. Some species are particularly suited for exploiting recently disturbed sites. Vegetation with the potential for rapid growth can quickly take advantage of the lack of competition. In the northeastern United States, shade-intolerant trees like pin cherry and aspen quickly fill in forest gaps created by fire or windthrow (or human disturbance). Silver maple and eastern sycamore are similarly well adapted to floodplains. They are highly tolerant of standing water and will frequently dominate floodplains where other species are periodically wiped out.

Forest fire burns on the island of Zakynthos in Greece on July 25th, 2007.

Another species which is well adapted to a particular disturbance is the Jack Pine in boreal forests exposed to crown fires. They, as well as some other pine species, have specialized serotinous cones that only open and disperse seeds with sufficient heat generated by fire. As a result, this species often dominates in areas where competition has been reduced by fire.

Species that are well adapted for exploiting disturbance sites are referred to as pioneers or early successional species. These shade-intolerant species are able to photosynthesize at high rates and as a result grow quickly. Their fast growth is usually balanced by short life spans. Furthermore, although these species often dominate immediately following a disturbance, they are unable to compete with shade-tolerant species later on and replaced by these species through succession.

While plants must deal directly with disturbances, many animals are not as immediately affected by them. Most can successfully evade fires, and many thrive afterwards on abundant new growth on the forest floor. New conditions support a wider variety of plants, often rich in nutrients compared to pre-disturbance vegetation. The plants in turn support a variety of wildlife, temporarily increasing biological diversity in the forest.

Importance

Biological diversity is dependent on natural disturbance. The success of a wide range of species from all taxonomic groups is closely tied to natural disturbance events such as fire, flooding, and windstorm. As an example, many shade-intolerant plant species rely on disturbances for successful establishment and to limit competition. Without this perpetual thinning, diversity of forest flora can decline, affecting animals dependent on those plants as well.

A good example of this role of disturbance is in ponderosa pine (*Pinus ponderosa*) forests in the western United States, where surface fires frequently thin existing vegetation allowing for new growth. If fire is suppressed, douglas fir (*Pesudotsuga menziesii*), a shade tolerant species, eventually replaces the pines. Douglas firs, having dense crowns, severely limit the amount of sunlight reaching the forest floor. Without sufficient light new growth is severely limited. As the diversity of surface plants decreases, animal species that rely on them diminish as well. Fire, in this case, is important not only to the species directly affected but also to many other organisms whose survival depends on those key plants.

References

- Soule, Michael E. (1986). Conservation Biology: The Science of Scarcity and Diversity. Sinauer Associates. p. 584. ISBN 978-0-87893-795-0.
- Meffe, Gary K.; Martha J. Groom (2006). Principles of conservation biology (3rd ed.). Sunderland, Mass: Sinauer Associates. ISBN 0-87893-518-5.
- van Dyke, Fred (2008). Conservation Biology: Foundations, Concepts, Applications, 2nd ed. Springer Verlag. p. 478. ISBN 978-1-4020-6890-4.
- Wilson, David Alec (2002). Darwin's cathedral: evolution, religion, and the nature of society. Chicago: University of Chicago Press. ISBN 0-226-90134-3.
- Primack, Richard B. (2004). A Primer of Conservation Biology, 3rd ed. Sinauer Associates. pp. 320pp. ISBN 0-87893-728-5.
- Davis, Peter (1996). Museums and the natural environment: the role of natural history museums in biological conservation. London: Leicester University Press. ISBN 0-7185-1548-X.
- Dyke, Fred Van (2008-04-07). "Conservation biology: Foundations, concepts, applications". ISBN 978-1-4020-6890-4.
- Gore, Albert (1992). Earth in the balance: ecology and the human spirit. Boston: Houghton Mifflin. ISBN 0-395-57821-3.
- MacArthur, R.H.; Wilson, E.O. (2001). The Theory of Island Biogeography. Princeton, N.J: Princeton University Press. ISBN 0-691-08836-5.
- European Communities (2008). The economics of ecosystems and biodiversity. Interim Report (PDF). Wesseling, Germany: Welzel+Hardt. ISBN 978-92-79-08960-2.
- Carroll, C. Dennis; Meffe, Gary K. (1997). Principles of conservation biology. Sunderland, Mass: Sinauer. ISBN 0-87893-521-5.
- Ehrlich, Anne H.; Ehrlich, Paul R. (1981). Extinction: the causes and consequences of the disappearance of species. New York: Random House. ISBN 0-394-51312-6.
- America has been the context for both the origins of conservation history and its modern form, environmental history. Asiaticsociety.org.bd. Retrieved on 2011-09-01.

- America has been the context for both the origins of conservation history and its modern form, environmental history. Asiaticsociety.org.bd. Retrieved on 2011-09-01.

Permissions

All chapters in this book are published with permission under the Creative Commons Attribution Share Alike License or equivalent. Every chapter published in this book has been scrutinized by our experts. Their significance has been extensively debated. The topics covered herein carry significant information for a comprehensive understanding. They may even be implemented as practical applications or may be referred to as a beginning point for further studies.

We would like to thank the editorial team for lending their expertise to make the book truly unique. They have played a crucial role in the development of this book. Without their invaluable contributions this book wouldn't have been possible. They have made vital efforts to compile up to date information on the varied aspects of this subject to make this book a valuable addition to the collection of many professionals and students.

This book was conceptualized with the vision of imparting up-to-date and integrated information in this field. To ensure the same, a matchless editorial board was set up. Every individual on the board went through rigorous rounds of assessment to prove their worth. After which they invested a large part of their time researching and compiling the most relevant data for our readers.

The editorial board has been involved in producing this book since its inception. They have spent rigorous hours researching and exploring the diverse topics which have resulted in the successful publishing of this book. They have passed on their knowledge of decades through this book. To expedite this challenging task, the publisher supported the team at every step. A small team of assistant editors was also appointed to further simplify the editing procedure and attain best results for the readers.

Apart from the editorial board, the designing team has also invested a significant amount of their time in understanding the subject and creating the most relevant covers. They scrutinized every image to scout for the most suitable representation of the subject and create an appropriate cover for the book.

The publishing team has been an ardent support to the editorial, designing and production team. Their endless efforts to recruit the best for this project, has resulted in the accomplishment of this book. They are a veteran in the field of academics and their pool of knowledge is as vast as their experience in printing. Their expertise and guidance has proved useful at every step. Their uncompromising quality standards have made this book an exceptional effort. Their encouragement from time to time has been an inspiration for everyone.

The publisher and the editorial board hope that this book will prove to be a valuable piece of knowledge for students, practitioners and scholars across the globe.

Index